Advances in Machining and Manufacturing Technology

Advances in Machining and Manufacturing Technology

Contributors

Ranbir Singh, Rajender Singh et al.

AURIS
Reference

www.aurisreference.com

Advances in Machining and Manufacturing Technology

Contributors: Ranbir Singh, Rajender Singh et al.

Published by Auris Reference Limited

www.aurisreference.com

United Kingdom

Advances in Machining and Manufacturing Technology

ISBN: 978-1-78154-931-5

British Library Cataloguing in Publication Data
A CIP record for this book is available from the British Library

Printed in the United Kingdom

Exclusively distributed by CBS Publishers & Distributors Pvt. Ltd.

Sales & Distribution Rights only for India, Pakistan, Bangladesh, Sri Lanka, Nepal and Bhutan.This book is not to be sold outside these territories.

Contents

List of Abbreviations

ACC	Adaptive Control with Constraints
ANFIS	Adaptive neurofuzzy inference system
ACO	Ant Colony Optimization
ANN	Artificial neural network
CVD	Chemical vapor deposition
CML	Combined machine loading
CNC	Computer numerical control
DE	Design Exploration
DOF	Degrees of Freedom
EDA	Electrical discharge alloying
EDM	discharge machining
EDM	Electrical discharge machining
FEA	Finite Element Analysis
FMS	Flexible Manufacturing System
GEP	Genetic expression programming
GHG	Greenhouse gas
ICDR	International Center for Dielectric Research
LIBWE	Laser-induced back-side wet etching
LCS	Learning Classifier System
LMP	Linear Mathematical Programming
LPCVD	Low pressure chemical vapour deposition
MRR	Material removed ratio
MRE	Mean relative errors
MIP	Mixed Integer Programming
NSERC	Natural Science and Engineering Research Council of Canada
NSFC	Natural Science Foundation of China
PSO	Particle Swarm Optimization
PVD	Physical vapor deposition
PVDF	Polyvinylidene fluoride
QCM	Quartz crystal micro-balance
RBFNs	Radial basis function neural networks
VCM	Varying-coefficient model
XRD	X-ray diffraction

List of Contributors

Ranbir Singh
Research Scholar, Department of Mechanical Engineering, DCRUST Murthal, Sonepat, India

Rajender Singh
Professor, Department of Mechanical Engineering, DCRUST Murthal, Sonepat, India

B. K. Khan
Technical Advisor, MSIT, Sonepat, India

Xiaomeng Li
Electronic Materials Research Laboratory, Key Laboratory of the Ministry of Education & International Center for Dielectric Research, Xi'an Jiaotong University, Xi'an 710049, China

Xiaoqing Wu
Electronic Materials Research Laboratory, Key Laboratory of the Ministry of Education & International Center for Dielectric Research, Xi'an Jiaotong University, Xi'an 710049, China

Peng Shi
Electronic Materials Research Laboratory, Key Laboratory of the Ministry of Education & International Center for Dielectric Research, Xi'an Jiaotong University, Xi'an 710049, China

Zuo-Guang Ye
Electronic Materials Research Laboratory, Key Laboratory of the Ministry of Education & International Center for Dielectric Research, Xi'an Jiaotong University, Xi'an 710049, China
Department of Chemistry and 4D LABS, Simon Fraser University, Burnaby, BC V5A 1S6, Canada

Qiaokang Liang
College of Electric and Information Technology, Hunan University, Changsha 410082, Hunan, China
State Key Laboratory of Advanced Design and Manufacturing for Vehicle Body, Hunan University, Changsha 410082, Hunan, China

Dan Zhang
Lassonde School of Engineering, York University, Toronto, ON M3J 1P3, Canada

Gianmarc Coppola
Lassonde School of Engineering, York University, Toronto, ON M3J 1P3, Canada

Jianxu Mao
College of Electric and Information Technology, Hunan University, Changsha 410082, Hunan, China

Wei Sun
College of Electric and Information Technology, Hunan University, Changsha 410082, Hunan, China
State Key Laboratory of Advanced Design and Manufacturing for Vehicle Body, Hunan University, Changsha 410082, Hunan, China

Yaonan Wang
College of Electric and Information Technology, Hunan University, Changsha 410082, Hunan, China
State Key Laboratory of Advanced Design and Manufacturing for Vehicle Body, Hunan University, Changsha 410082, Hunan, China

Yunjian Ge
Institute of Intelligent Machines, Chinese Academy of Science, Hefei 230031, Anhui, China

Rosario Domingo
Departamento de Ingeniería de Construcción y Fabricación, Universidad Nacional de Educación a Distancia (UNED), C/Juan del Rosal 12, Madrid 28040, Spain

Marta María Marín
Departamento de Ingeniería de Construcción y Fabricación, Universidad Nacional de Educación a Distancia (UNED), C/Juan del Rosal 12, Madrid 28040, Spain

Juan Claver
Departamento de Ingeniería de Construcción y Fabricación, Universidad Nacional de Educación a Distancia (UNED), C/Juan del Rosal 12, Madrid 28040, Spain

Roque Calvo
Departamento de Ingeniería Mecánica, Química y Diseño Industrial, Universidad Politécnica de Madrid, Ronda de Valencia 3, Madrid 28012, Spain

Wan-Jui Shen
Department of Materials Science and Engineering, National Tsing Hua University, Hsinchu 30013, Taiwan

Ming-Hung Tsai
Department of Materials Science and Engineering, National Chung Hsing University, Taichung 40227, Taiwan

Jien-Wei Yeh
Department of Materials Science and Engineering, National Tsing Hua University, Hsinchu 30013, Taiwan

L. Rihakova
Palacky University, RCPTM, Joint Laboratory of Optics of Palacky University and Institute of Physics of the Academy of Sciences of the Czech Republic, 17. Listopadu 50a, 77207 Olomouc, Czech Republic

H. Chmelickova
Institute of Physics of the Academy of Sciences of the Czech Republic, Joint Laboratory of Optics of Palacky University and Institute of Physics of the Academy of Sciences of the Czech Republic, 17. Listopadu 50a, 77207 Olomouc, Czech Republic

Wei Feng
Department of Mechanical and Electrical Engineering, School of Physics and Mechanical & Electrical Engineering, Xiamen University, Xiamen 361005, China

Bin Yao
Department of Mechanical and Electrical Engineering, School of Physics and Mechanical & Electrical Engineering, Xiamen University, Xiamen 361005, China

BinQiang Chen
State Key Laboratory for Manufacturing and Systems Engineering, School of Mechanical Engineering, Xi'an Jiaotong University, Xi'an 710049, China

DongSheng Zhang
School of Mechanical Engineering, Shaanxi University of Technology, Hanzhong 723001, China

XiangLei Zhang
Department of Mechanical and Electrical Engineering, School of Physics and Mechanical & Electrical Engineering, Xiamen University, Xiamen 361005, China

ZhiHuang Shen
Department of Mechanical and Electrical Engineering, School of Physics and Mechanical & Electrical Engineering, Xiamen University, Xiamen 361005, China

Mi Xiao
State Key Laboratory of Digital Manufacturing Equipment and Technology, Huazhong University of Science and Technology, Wuhan 430074, China

Long Wen
State Key Laboratory of Digital Manufacturing Equipment and Technology, Huazhong University of Science and Technology, Wuhan 430074, China

Xi Li
School of Mechanical Science and Engineering, Huazhong University of Science and Technology, Wuhan 430074, China

Liang Gao
State Key Laboratory of Digital Manufacturing Equipment and Technology, Huazhong University of Science and Technology, Wuhan 430074, China

Agnieszka Dmowska Bogdan Nowicki
Institute of Manufacturing Technology, Warsaw University of Technology, 02-524 Warsaw, Poland

Anna Podolak-Lejtas
Institute of Manufacturing Technology, Warsaw University of Technology, 02-524 Warsaw, Poland

Vivek Jain Apurbba Kumar Sharma
Mechanical and Industrial Engineering Department, Indian Institute of Technology Roorkee, Roorkee, Uttarakhand 247667, India

Pradeep Kumar
Mechanical and Industrial Engineering Department, Indian Institute of Technology Roorkee, Roorkee, Uttarakhand 247667, India

S. Belmokhtar, A.I. Bratcu
University Campus STeP Ri Slavka Krautzeka 83/A 51000 Rijeka, Croatia

A. Dolgui
University Campus STeP Ri Slavka Krautzeka 83/A 51000 Rijeka, Croatia

Gokula Vijaykumar Annamalai Vasantha
Design, Manufacture and Engineering Management, University of Strathclyde, Glasgow G1 1XQ, UK

Hitoshi Komoto
Advanced Manufacturing Research Institute, National Institute of Advanced Industrial Science and Technology, Tsukuba, Ibaraki 305-8564, Japan

Romana Hussain
School of Engineering, Cranfield University, Cranfield, Bedfordshire MK43 0AL, UK

Rajkumar Roy
Manufacturing and Materials Department, Cranfield University, Cranfield, Bedfordshire MK43 0AL, UK

Tetsuo Tomiyama
Manufacturing and Materials Department, Cranfield University, Cranfield, Bedfordshire MK43 0AL, UK

Steve Evans
Institute for Manufacturing, University of Cambridge, Cambridge CB3 0FS, UK

Ashutosh Tiwari
Manufacturing and Materials Department, Cranfield University, Cranfield, Bedfordshire MK43 0AL, UK

Stewart Williams
Welding Engineering and Laser Processing Centre, Cranfield University, Cranfield, Bedfordshire MK43 0AL, UK

Preface

Manufacturing technology provides the tools that enable production of all manufactured goods. Production tools include machine tools and other related equipment and their accessories and tooling. The text *Advances in Machining and Manufacturing Technology* presents most recent advances in the fields of machining and advanced manufacturing technology. It describes both time-tested and modern methods of manufacturing engineering materials. First chapter presents a research review of the optimization techniques and the objectives for which the machine loading problem in flexible manufacturing system (FMS) has been solved, and scope of the research in the field. In second chapter, we discuss a new approach to the fabrication of integrated silicon-based piezoelectric diaphragm-type biosensors by using sodium potassium niobate-silver niobate composite lead-free thin film as the piezoelectric layer. Third chapter focuses on design and analysis of a sensor system for cutting force measurement in machining processes. Fourth chapter analyzes the influence of the cutting conditions and the characteristics of cutting tools on the energy required in machining processes and the carbon dioxide equivalent (CO2-eq) emissions generated per material removed ratio (MRR) in an effort to define common criteria for using cutting inserts in a sustainable manner. Machining performance of sputter-deposited $(Al_{0.34}Cr_{0.22}Nb_{0.11}Si_{0.11}Ti_{0.22})_{50}N_{50}$ high-entropy nitride coatings has been investigated in fifth chapter. In sixth chapter, we report the mechanisms of laser micromachining of materials. Seventh chapter provides an integrated grinding model combining the machine and process models, which can be used to predict process-machine interactions in grinding process. Eighth chapter focuses on the transient current of the stable milling process and established a predictive model for the transient current amplitude. Ninth chapter presents the results of the influence of basic electrical discharge machining (EDM) parameters and electrical discharge alloying (EDA) parameters on surface layer properties and on selected performance properties of machine parts. Tenth chapter provides an overview of the main issues concerning different aspects of micro-USM, its performance, and limitations in the application. Eleventh chapter deals with the designing of modular reconfigurable transfer lines, where a set of standard spindle heads are used to produce a family of similar products. A manufacturing framework for capability-based product-service systems design has been proposed in last chapter.

Chapter 1

A CRITICAL REVIEW OF MACHINE LOADING PROBLEM IN FLEXIBLE MANUFACTURING SYSTEM

Ranbir Singh[1], Rajender Singh[2], B. K. Khan[3]

[1]Research Scholar, Department of Mechanical Engineering, DCRUST Murthal, Sonepat, India

[2]Professor, Department of Mechanical Engineering, DCRUST Murthal, Sonepat, India

[3]Technical Advisor, MSIT, Sonepat, India

ABSTRACT

Production planning is the foremost task for manufacturing firms to deal with, especially adopting Flexible Manufacturing System (FMS) as the manufacturing strategy for production seeking an optimal balance between productivity-flexibility requirements. Production planning in FMS provides a solution to problems regarding part type selection: machine grouping, production ratio, resource allocation and loading problem. These problems need to be solved optimally for maximum utilization of resources. Optimal solution to these problems has been a focus of attention in production and manufacturing, industrial and academic research since a number of decades. Evolution of new optimization techniques, software, technology, machines and computer languages provides the scope of a better optimal solution to the existing problems. Thus there remains a need of research to solve the problem with latest tools and techniques for higher optimal use of available resources. As an objective, the researchers need to reduce the computational time and cost, complexity of the problem, solution approach viz. general or customized, better user friendly communication with machine, higher freedom to select the desired objective(s) type(s) for optimal solution to the problem. As an approach to the solution to the problem, a researcher first needs to go for an exhaustive literature review, where the researcher needs to find the research gaps, compare and analyze the tools and techniques used, number of objectives considered for optimization and need, and scope of research for the research problem. The present study is a review paper analyzing the research gaps, approach and techniques used, scope of new optimization techniques or any other research, objectives considered and validation approaches for loading problems of production planning in FMS.

INTRODUCTION

Manufacturing is the pilot element within the overall enterprise. Possible manufacturing outputs of the firm to meet pre-determined corporate level goals should be known to remain in competition at global market. Manufacturing strategy writes the script to calculate possible manufacturing outputs. Existence of the manufacturing strategy guides daily decisions and activities with clear understanding of decision-goal relationship of the corporation and provides a vision for the firm to remain aligned with the overall business strategy of the firm. The firms having manufacturing strategies for achieving corporate goals survive for long run. A strategy is also a strong communication tool between different levels of management to bring all operations in line with corporate objectives. Custom manufacturing, continuous manufacturing, intermittent manufacturing, flexible manufacturing, just-in-time manufacturing, lean manufacturing and agile manufacturing are major manufacturing strategies revealed in the literature.

FMS is an automated manufacturing system consisting of computer numerical control (CNC) machines with automated material handling, storage and retrieval system. The aim of FMS is to attain the efficiency of mass production while utilizing the flexibility of job shop simultaneously. FMS is adopted for batch production of mid production volume and mid part variety (flexibility) requirements. Since its evolution, researchers are working for optimality of FMS strategy. FMS is a field of great potential hence a numerous complex planning problems need to be solved. Major complex production planning problems are part type selection: machine grouping, production ratio, resource allocation and loading problem (Stecke, 1983). All the production planning problems need to be optimally solved. The present research is the critical literature review for the loading problem of production planning in FMS.

Tooling individual or group of machine(s) to collectively accomplish all manufacturing operations concurrently for all part type in a batch is termed as loading problem. A solution to the problem specifies the machine(s) to which a job has to be routed in sequence for each of its operation(s) with respective tooling under capacity and technological constraint(s) for all jobs in a batch simultaneously to achieve certain objective(s). Loading is a complex combinational planning problem because a batch of jobs is to be machined simultaneously and each job requires unique set of operations effect on manufacturing cost. To solve the problem, highly experienced and skilled professionals are required. Without the use of some computational or optimization technique, the solution may or may not be optimal. Thus there arises the need of optimal solution with the help of computational methods using optimization techniques. The paper is a critical review paper analyzing the research gaps, approach and techniques used, scope of new optimization techniques or any other research, objectives considered and validation approaches for loading problems of production planning in FMS.

LITERATURE REVIEW OF LOADING PROBLEMS IN FMS

In brief, to solve a problem using optimization techniques and computational analysis, objective(s) are first set, the physical system is modelled using certain technique like mathematical modelling, the solution is then derived under given boundary conditions and constraints to achieve the given objectives, the results are then analysed and the solution approach is then validated. Heuristics has been widely used by the researchers. Table 1 presents the tabulated research review discussing the approach, objectives and results of the loading problems in FMS. Flexible manufacturing is an overall pilot element within an enterprise. Each multinational manufacturing concern has to satisfy business goals to remain in competition with the global market. The manufacturing firm should be aware of the possible manufacturing outputs that will closely match the goals and strategy determined at the corporate level. The existence of a manufacturing strategies guide the daily decisions and activities with clear understanding of how those daily decisions relate to the overall goals of the corporation. The firms having manufacturing strategies for achieving corporate goals survive long. A manufacturing strategy provides a vision to the manufacturing organization for keeping itself aligned with the overall business strategy of the corporation.

A strategy is also a strong communication tool between different levels of management to bring all operations in line with corporate objectives. Custom manufacturing, continuous manufacturing, intermittent manufacturing, flexible manufacturing, just-in-time manufacturing, lean manufacturing and agile manufacturing are the major manufacturing strategies which are revealed in literature.

FMS is an automated manufacturing system consisting of numerical control (computer) machines with automated material handling, automated storage and retrieval system. The aim of FMS is to attain the efficiency of mass production while utilizing the flexibility of job shop simultaneously.

Table 1: Review of machine loading problems in FMS based on heuristics approach

			HEURISTIC APPROACH		
Sr.	Year	Researcher Name	Approach	Objectives	Results	Validation approach
1	1983	K. E. Stecke & F. Brian Talbot [2]	Heuristic methods	Ø Minimizing part movements Ø Balancing of workload Ø Unbalancing of workload	Determined how machine tool magazine in a FMS can be loaded to meet simultaneous requirements of a number of different parts	Computational results are presented
2	1985	K. Shankar & Y. J. Tzen [14]	Heuristic methods	Ø Minimizing system unbalance Ø Number of late jobs Ø Balancing of workload	Computational results presented gives improved results	performance is compared with previous results from literature
3	1988	J. A. Ventura, F. F. Chen, & M. S. Leonard [15]	Heuristic algorithms	Ø Minimizing make span	Improved Performance	The performance of each of the proposed algorithms is evaluated by testing on two hypothetical FMSs.
4	1990	B. Ram, S. Sarin, & C. S. Chen [16]	Fast heuristic algorithms	Ø Maximizing throughput	FMS loading problem can be solved near optimally in short time	Computational results are produced and compared with previous results
5	1992	S. K. Mukhopadhyay, S. Midha, & V. Murlikrishna [17]	Heuristic procedure	Ø Minimizing system unbalance Ø Maximizing throughput	Results show that algo developed is very reliable and efficient	tested on ten problems and are compared with existing results
6	1993	K. Kato, F. Oba, & F. Hashimoto [18]	Heuristic approach	Ø Minimizing total number of cutting tools required Ø Maximizing utilization rate of each machine	Computational results shows improved effectiveness	Computational results are given to demonstrate the effectiveness of the proposed method
7	1995	E. K. Stecke & F. Brian Talbot [19]	Heuristic Algorithms	Ø Minimizing part movements Ø balancing of workload Ø unbalancing of workload	Results are computationally demonstrated & found improved significantly	Computational results are produced and compared with previous results
8	1997	M. K. Tiwari et al. [20]	Heuristic solution approach	Ø Maximizing throughput Ø Minimizing system unbalance	Graphical representation and subsequent model validation	Computational results are produced and compared with previous results
9	1998	G. K. Nayak & D. Acharya [21]	Heuristics and mathematical programming approaches	Ø Maximizing part types in each batch Ø Maximizing routing flexibility of batches	Heuristic proposed for part type selection & simple mathematical programs for other two problems	Computational results are compared with existing results
10	2000	D.-H. Lee & Y.-D. Kim [22]	Heuristic algorithms	Ø Minimizing maximum workload of machines	Results show that suggested algos perform better than existing	Simulation results are compared with existing results
11	2006	N. Nagarjunaa, O. Maheshb, & K. Rajagopal [23]	Heuristic based on multi stage programming approach	Ø Minimizing system unbalance	Bring together productivity of flow lines and flexibility of job shops	Tested on 10 sample problems available in FMS literature and compared with existing solution methods
12	2006	M. Goswami & M. K. Tiwari [24]	Heuristic-based approach	Ø Minimizing system unbalance Ø Maximizing throughput	Loading problem is crucial link between tactical planning and operational decisions	Extensive computational experiments have been carried out to assess the performance of the proposed heuristic and validate its relevance
13	2007	M. K. Tiwari, J. Saha, & S. K. Mukhopadhyay [25]	Heuristic Solution Approaches	Ø Minimizing system unbalance Ø Maximizing throughput	GA based heuristic are found more efficient and outperform in terms of solution quality	tested on problems representing three different FMS scenarios from available literature

Most of the researches are focused on increasing the production volume of FMS with increased part varieties. FMS is an interesting field of research to solve the issues and problems encountered by industries. Though FMS has great potential benefits, a numerous control and planning problem need to be taken care of. Kathryn E. Stecke in 1983 described five complex productions planning problems namely part type selection problem, Machine Grouping Problem, production ratio problem, resource allocation problem and loading problem [1] .

Loading means allocation of the operations and required tools to a part types among the set of machine(s), subjected to resource & technological constraints to collectively accomplish all manufacturing operations for each pat type machined concurrently. The allocation of workloads to the existing production facilities for manufacturing products with several constraints in order to perform production activities according to the production plan established, it is essential to adjust the workload for each of the facilities and workers in each time period so they are not assigned work exceeding the given capacity. A solution to this problem specifies the tools which must be loaded in each machine tool magazine and the machine(s) to which a part can be routed for each of its operations before production begins. A variety of products are manufactured simultaneously in FMS, where each part requires potentially unique set of operations, and loading problem is declared as a combinational problem by Kathryn E. Stecke [2] which is highly complex, time-consuming and tedious in nature & requires highly experienced process planners.

Machine loading is one of the most critical production planning problems of FMS. It concerns with the time spend by the job(s) on machine(s) and the manufacturing cost. Manufacturing cost is the sum of fixed and variable costs. Variable cost varies with the level of production output. As output increases, variable cost increases. Once invested, we can't play around the fixed cost; hence to reduce the manufacturing cost, researcher has to minimize the variable cost while maximizing the output. This is done by developing and optimizing a virtual model of manufacturing by some conventional or non-conventional technique for certain number of objectives with their individual weightage accordingly. A researcher has to solve the manufacturing model to minimize the time spent by the job on machine, number of tool used, and movements of tool and job. FMS is a group technology concept hence all the operations on the group jobs are required to be completed at once keeping in view that no machine should be idle or overloaded at any instance of time. Thus the optimized solution of the machine loading problem for certain objectives under technological and capacity constraints is required. The solution to the machine loading problem is to minimize the manufacturing cost as a whole. Increasing part varieties with raised productivity is necessary to be in competition and to maintain the demand of the product, which is possible by continuous research and optimized solutions to each of the production planning problem. This paper presents a research review of the optimization techniques and the objectives for which the machine loading problem in FMS has been solved, and scope of the research in the field.

Before presenting literature review, an introduction to the optimization techniques and their classification seems necessary to be discussed here for better understanding of the subject. Optimization is the approach for ideal solution. Accuracy of the solution depends on the approach, modeling, computational time and capacity, and nature of the problem. Optimization is classified into six categories: function, and trial and error, single variable and multiple variables, static and dynamics, continuous and discrete, constrained and unconstrained, and random and minimum seeking.

A functional optimization is for theoretical approach where a mathematical formula describes the objective function. Trial-and-error optimization is for experimental optimization with change in the variables which affect output without knowing much about the process. An optimization can be single variable for one dimensional analysis, and multi variables for multi-dimensional analysis. As the number of variables increases, the complexity of the problem also increases. Static optimization is independent of time and dynamic optimization as a function of time. Discrete optimization has a finite number of variables with all possible values, while continuous optimization has infinite number of variables with all possible values. Values are incorporated in equalities and inequalities to an objective of variable function in constrained optimization while the variables can take any value in unconstrained optimization. Random optimization finds sets of variables by probabilistic calculations while minimal seeking is the traditional optimization algorithms which are generally based on calculus methods and minimizes the function by starting from an initial set of variable values.

These optimization approaches can be further sub-categorized as stochastic programming, integer programming, linear programming, nonlinear programming, bound programming, network programming, least squares methods, global optimization, and non-differential optimization.

Most of the researches are focused on solving the machine loading problem by global optimization algorithms. Global optimization algorithms are generally categorized into two approaches: deterministic and probabilistic. Deterministic are sub-categorized into static space search (1992) [3] , branch and bond and algebraic geometry algorithms. Probabilistic is sub-categorized as Monte Carlo algorithms, soft computing and Artificial Intelligence (AI). Monte Carlo algorithms includes two classes, one covers Stochastic (hill climbing) (2002) [4] , Random optimization (1963) [5] , Simulated Annealing (SA) (1953) [6] , Tabu Search (TS) (1989) [7] , Parallel tempering, Stochastic tunneling and Direct Monte Carlo Sampling, and second class includes Evolutionary Computation (EC). EC can be performed by Monte Carlo algorithms or soft computing or AI. EC is further classified as Evolutionary Algorithms (EA), Memetic Algorithms (hybrid Algorithms) (1989) [8] , Harmonic Search (HS), Swarm Intelligence (SI). EA is sub-classified as Genetic Algorithms (GA) (1962) [9] , Learning Classifier System (LCS) (1977) [10] , Evolutionary Programming, Evolution Strategy (ES), Genetic Programming (GP) (1958) [11] . ES includes Differential Evolution (DE), and GP includes Standard GP, Linear GP and Grammar Guided GP. SI includes Ant Colony Optimization (ACO) (1996) [12] and Particle Swarm Optimization (PSO) (1995) [13]. The above discussed classifications scheme will be used for classifying the optimization techniques for solving the machine loading problems of FMS in the paper. Figure 1 shows the evolution of the major optimization techniques along the time axis.

LITERATURE REVIEW OF MACHINE LOADING PROBLEMS IN FMS

An exhaustive research review has been carried out for study of approaches and optimization techniques for machine loading problems in FMS. A. Baveja, A. Jain, A. K. Singh, A. Kumar, A. M. Abazari, A. Murthy, A. Prakash, A. Srinivasulu, A. Turkcan, C. A. Yano, C. Basnet , C.S. Chen, D. Acharya, D. Kosucuoglu, D.H. Lee, F. Brian Talbot, F. F. Chen, F. Guerrero, F. Hashimoto, F. Oba, G. K. Nayak, G.C. Lee, H. C. Co, H. Sattari, H. Yong, H.B. Jun, H.-K. Roh, J. A. Ventura, J. Larranaeta, J. S. Bicrmann, J. Saha, J. G. Shanthikumar, J. N. D. Gupta, K Chandrashekara, K. E. Stecke, K. Kato, K. M. Bretthauer, K. Rajagopal, K. Shankar, L. H. S. Luong, L. S. Kiat, M. A. Gamila, M. A. Venkataramanan, M. Arıkan, M. Berrada, M. Goswami, M. I. Mgwatua, M. K. Pandey, M. K. Tiwari, Ming Liang, M. M. Aldaihani, M. S. Akturk, M. S. Leonard, M. Savsar, M. Solimanpur, M. Yogeswaran, N. K. Vidyarthi, N. Khilwani, N. Kumar, N. Nagarjunaa, N. K. Vidyarthi, O. Maheshb, Prakash, R. P. Sadowski, R. Budiarto, R. D. Matta, R. H. Storer, R. M. Marian, R. R. Kumar, R. Shankar, R. Swarnkar, S. Biswas, S. Deris, S. Erol, S. G. Ponnambalam, S. K. Mandal, S. K. Mukhopadhyay, S. Kumar, S. Lozano, S. Midha, S. Motavalli, S. P. Dutt, S. Rahimifard, S. S. Mahapatra, S.C. Sarin, S.K. Chen, S.K. Lim, S.T. Newman, T. J. Greene, T. J. Sawik, T. Koltai, T. L. Morin, T. Sawik, U. Bilge, U. K. Yusof, V. H. Nguyen, V. M Kumar, V. Murlikrishna, V. N. Hsu, V. Tyagi, W. F. Mahmudy, Y. Cohen, Y. D. Kim, Y. J. Tzen and Z. Wu are key researchers for solving the loading problem of production planning in FMS. The tabulated research review discussing the approach, objectives, results and validation approach for machine loading problems in FMS is discussed in Tables 1-3. The literature review is classified into three groups: (1) heuristics; (2) global optimization; and (3) other optimization techniques.

Table 1 presents the review of machine loading problems of FMS based on heuristics approach. The heuristics approach has been significantly used for solving the research problem. Research has gained significant acceleration with the evolution and growth of global optimization techniques.

Table 2 presents the review of machine loading problems in FMS based on global optimization algorithms. Global optimization techniques have been explored rigorously by the researchers. The natural selection techniques have reported good results compared to others. The application of global optimization techniques for solving machine loading problem is increasing with growth of natural optimization techniques. The results reported by natural optimization techniques are more acceptable. Natural optimization techniques, GA and PSO are widely used techniques.

Table 3 presents the review of machine loading problems in FMS based on optimization techniques not falling in the above classification. Since the major focus is on heuristics and global optimization techniques, thus other techniques are grouped in a single table. These techniques have been adopted from time to time for solving the machine loading problem as shown year wise in Table 3.

Optimization techniques and approaches under the classification of global optimization scheme are discussed in Table 2.

Optimization techniques and approaches not falling under the above classifications are discussed in Table 3.

Table 4 has been formulated on regressive analysis of Tables 1-3, for the analysis of the loading objectives to be fulfilled while solving the loading problem. It is a year-wise tabulation and analysis of the loading objectives.

Table 2: Review of machine loading problems in FMS based on global optimization algorithms

Sr.	Year	Researcher Name	Approach	Objectives	Results	Validation approach
			GLOBAL OPTIMIZATION ALGORITHMS			
			a. Deterministic approach			
			1. Branch and bound			
1	1986	M. Berrada & K. E. Stecke [26]	Branch and bound approach	➤ Balancing of workload	Computational results gives fruitful results	Computational results are produced and demonstrated the efficiency of suggested procedures
2	1989	K. Shankar & A. Srinivasulu [27]	Branch & backtrack procedure and Heuristic procedures	➤ Maximizing assigned workload ➤ Maximizing throughput ➤ Minimizing workload unbalance	Each procedure is illustrative by numerical example and results are with improved performance	An illustrative numerical example
3	1994	Y. D. Kim & C. A. Yano [28]	New branch and bond algorithm	➤ Maximizing throughput	Improved efficiency	Computational results are produced and compared with previous results
			2. Algebraic Geometry			
4	1986	T. J. Greene & R. P. Sadowski [29]	Mixed integer programming	➤ Minimizing make span ➤ Minimizing mean flow time ➤ Minimizing mean lateness	Explained simple numeric example	a simple numeric example
5	1987	S.C. Sarin & C.S. Chen [30]	Mathematical model	➤ Minimizing overall machining cost	Computational results are reported	Computational results are compared with literature results
6	1990	K. M. Bretthauer & M. A. Venkataramanan [31]	Linear Integer Programming	➤ Maximizing weighted sum of number of operation to machine assignments	Computational results are satisfactory with improved performance	Computational results are produced
7	1990	H. C. Co, J. S. Biermann, & S.K. Chen [32]	Mixed-integer programming (MIP)	➤ Balancing of workloads	Results were found practical	Computational results are produced
8	1990	M. Liang & S. P. Dutt [33]	Mixed-Integer Programming	➤ Minimizing production cost	Demand for change on optimal solution	An example problem is solved
9	1993	Ming Liang [34]	Non-linear programming	➤ Maximizing system output	production cost can be significantly reduced using this approach	Computational results with an illustrative example is demonstrated
10	1994	Ming Liang [35]	Non-linear programming	➤ Maximizing system output ➤ minimizing production cost	Production cost can be significantly reduced using this approach	An illustrative example is solved using the suggested approach
11	1997	V. N. Hsu & R. D. Matta [36]	Lagrangian-based heuristic procedure (MIP problem formulation)	➤ total processing cost	finds a good loading solution	iteratively compared different scenarios
12	1998	T. J. Sawik [37]	Integer programming & approximative lexicographic approach	➤ Balancing workloads ➤ Minimizing total interstation transfer time	Results of computational experiments are reported	illustrative example and some results of computational experiments
13	1999	F. Guerrero, S. Lozano, T. Koltai, & J. Larranaeta [38]	Mixed-integer linear program	➤ Balancing of workload	New approach to loading problem	Computational results are produced
14	2001	N. Kumar & K. Shanker [39]	Mixed integer programming	➤ Balancing of Workload	Results are in agreement with previous findings	Computational results are compared with the previous findings

15	2003	M. A. Gamila & S. Motavalli [40]	mixed integer programming	➤ Minimizing completion time ➤ Minimizing Material handling time ➤ Minimizing total processing time	Results reported increased efficiency and performance of system	Computational results are compared with the previous findings
16	2004	T. Sawik [41]	Mixed integer programming	➤ Minimizing production time	Computational results reported better performance	Numerical examples and some computational results are compared with available literature
17	2011	M. I. Mgwatua [42]	Linear Mathematical Programming	➤ Maximizing throughput ➤ Minimizing make span	More interactive decisions and well-balanced workload of the FMS can be achieved when sub-problems are solved jointly	Compared with results from previous literature
18	2012	A. M. Abazari, M. Solimanpur, & H. Sattari [43]	Linear mathematical programming	➤ Minimizing System unbalance	Genetic algorithm (GA) is proposed and performance of proposed GA is evaluated based on some benchmark problems	Performance is evaluated based on some benchmark problems adopted from the literature

b. Probabilistic

3. Monte Carlo algorithms

19	1998	S. K. Mukhopadhyay et al. [44]	Simulated annealing (SA) approach	➤ Minimizing system imbalance	Tried to give global optimum solution	Computational results are compared with existing results
20	2004	R. Swarnkar & M. K. Tiwari [45]	Hybrid tabu search and simulated annealing based heuristic approach	➤ Minimizing system unbalance ➤ Maximizing throughput	Results reported better performance	Tested on Standard problems and the results obtained are compared with those from some of the existing heuristics from literature
21	2005	M. M. Aldaihani & M. Savsar [46]	Stochastic model	➤ Minimizing total (FMC) flexible manufacturing cell cost per unit of production	Results reported better performance	Computational results were presented
22	2006	M. K. Tiwari, S. Kumar, S. Kumar, Prakash, & R. Shankar [47]	Constraints-Based Fast Simulated Annealing (SA) Algorithm	➤ Minimizing system unbalance ➤ Maximizing throughput	Proposed algorithm enjoys the merits of simple SA and simple genetic algorithm	The application of the algorithm is tested on standard data sets
23	2012	M. Arıkan & S. Erol [48]	Hybrid simulated annealing-tabu search algorithm	➤ Maximizing weighted sum ➤ Minimizing system unbalance ➤ Balancing of workload	Results shows improved system performance compared to earlier results in literature	The results are compared with those developed earlier by the authors

4. Evolutionary Computation (EC)

✓ Evolutionary algorithms (EA)

24	2000	N. Kumar & K. Shanker [49]	Genetic algorithm (GA)	➤ Maximizing number of part types in a batch ➤ Maximizing number of parts selected a batch ➤ Maximizing mean machine utilization	Results reported reduced computational requirements	comparative study of Computational results
25	2002	H. Yong & Z. Wu [50]	GA-based integrated approach	➤ Balancing of workloads	Results shows that suggested approach perform better than existing	Computational results are compared with the previous findings

No	Year	Author	Method	Objectives	Findings	Testing
26	2006	A. Kumar, Prakash, M. K. Tiwari, R. Shankar, & A. Baveja [51]	Constraint based genetic algorithm (CBGA)	➤ Balancing machine processing time ➤ Minimizing number of movements ➤ Balancing of workload ➤ Unbalancing of workload ➤ Filling the tool magazines as densely as possible ➤ Maximizing sum of operations priorities	The methodology developed here helps avoid getting trapped at local minima	The application of the algorithm is tested on standard data sets from available literature.
27	2007	A. Turkcan, M. S. Akturk, & R. H. Storer [52]	Genetic Algorithm (GA)	➤ Minimizing manufacturing cost ➤ Total weighted tardiness	Approach improves CNC machine efficiency & responsiveness to customer due date requirements	compared with the performance of most commonly used approach in the literature
28	2008	V. Tyagi & A. Jain [53]	Genetic algorithm based methodology	➤ Minimizing system unbalance	For a given number of tool copies of each tool type tool loading is affected by the availability of flexible process plans	An illustrative example
29	2012	U. K. Yusof, R. Budiarto, & S. Deris [54]	Constraint-chromosome genetic algorithm	➤ Minimizing system unbalance ➤ Maximizing throughput	Overall combined objective function increased by 3.60% from previous best result	tested on 10 sample problems available in the FMS literature and compared with existing solution methods

✓ Memetic (hybrid) Algorithms

No	Year	Author	Method	Objectives	Findings	Testing
30	2000	M. K. Tiwari & N. K. Vidyarthi [55]	Genetic Algorithm (GA) based (HA) Heuristic Approach	➤ Minimizing system unbalance ➤ Maximizing throughput	Optimal solution to problem	Tested on ten sample problems and the computational results obtained have been compared with those of existing methods
31	2009	M. Yogeswaran, S. G. Ponnambalam, & M. K. Tiwari [56]	Hybrid genetic algorithm simulated annealing algorithm (GASAA)	➤ Minimising system unbalance ➤ Maximising throughput	Results support better performance of GASA over algorithms reported in literature	results compared with reported in the literature
32	2010	S. K. Mandal, M. K. Pandey, & M. K. Tiwari [57]	Genetic algorithm simulated annealing Heuristics approach	➤ Minimizing breakdowns ➤ Minimizing system unbalance ➤ Minimizing make span ➤ Maximizing throughput	Results incurred under breakdowns validate robustness of developed model for dynamic ambient of FMS	Compared with dataset from previous literature
33	2012	V. M Kumar, A. Murthy, & K Chandrashekara [58]	Meta-hybrid heuristic technique based on genetic algorithm and particle swarm optimization	➤ Minimizing system unbalance ➤ Maximizing throughput	Model efficiency and performance of system is comparable with results compared to literature	Computational results are presented
34	2012	C. Basnet [59]	Hybrid genetic algorithm	➤ Minimizing system unbalance	Better solutions for system unbalance	Computational comparison between the genetic algorithm and previous algorithms is presented
35	2012	D. Kosucuoglu & U. Bilge [60]	Genetic algorithm based mathematical programming (GAMP)	➤ Minimizing total distance travelled by parts during production	GALP integration works successfully for this hard-to-solve problem	tested through extensive numerical experiments

✓ Swarm Optimization

No	Year	Author	Method	Objectives	Findings	Testing
36	2007	S. Biswas & S. S. Mahapatra [61]	Swarm Optimization Approach	➤ Minimizing system unbalance	Results reported improved system balance	compared with existing techniques for ten standard problems available in literature representing three different FMS scenarios

37	2008	S. Biswas & S. S. Mahapatra [62]	Modified particle swarm optimization	➤ Minimizing system unbalance	Proposed algorithm produces promising results in comparison to existing methods	comparison to existing methods for ten benchmark instances available in the FMS literature
38	2008	S. G. Ponnambalam & L. S. Kiat [63]	Particle Swarm Optimization (PSO)	➤ Minimizing system unbalance ➤ Maximizing throughput	Performance of PSO is satisfactory compared with heuristics reported in literature	tested by using 10 sample dataset and the results are compared with the heuristics reported in the literature
				✓ Artificial intelligence		
39	2001	N. K. Vidyarthi & M. K. Tiwari [64]	Fuzzy-based Heuristic Approach	➤ Minimizing system unbalance ➤ Maximizing throughput	Substantial improvement in solution quality over some existing heuristic-based approaches	Tested on 10 problems adopted from literatures and computational results are compared with the previous findings
40	2004	R. R. Kumar, A. K. Singh, & M.K. Tiwari [65]	Fuzzy based algorithm	➤ Minimizing system unbalance ➤ Maximizing throughput	Extended neuro fuzzy petri net is constructed	Computational results are compared with standard data set adopted from literature
41	2008	A. Prakash, N. Khilwani, M. K. Tiwari, & Y. Cohen [66]	Modified immune algorithm	➤ Maximizing throughput ➤ Minimizing system unbalance	Good results as compared to best results reported in literature	compared to the best results reported in the literature

The table is showing the list of objectives for which the loading problem is solved. The tick mark (√) in the table shows the density for repeatability of the objectives.

Abbreviations used in Table 4:

- Minimizing system unbalance
- Maximizing throughput
- Balancing of workload in the system configured of groups composed of machines of equal size
- Minimizing make span
- Meeting delivery dates
- Minimizing manufacturing cost/Minimizing total processing cost/ Minimizing total flexible manufacturing cell cost per unit of production
- Minimizing tardiness
- Minimizing production cost
- Unbalancing the workload per machine for a system of groups of pooled machines of unequal sizes
- Minimizing part movements
- Maximizing part types in each batch
- Minimizing subcontracting costs
- Maximizing weighted sum of number of operation to machine assignments
- Minimizing flow time
- Minimizing late jobs (number)/ lateness

- Minimizing machine processing time
- Minimizing production time
- Filling the tool magazines as densely as possible
- Maximizing assigned workload
- Maximizing routing flexibility of batches
- Maximizing the sum of operations priorities
- Minimizing material handling time
- Minimizing total distance travelled by parts during production
- Minimizing total number of cutting tools required
- Minimizing workload of machines
- Minimizing breakdowns
- Minimizing earliness

After regressive analysis of the loading objectives of various researchers the optimization approaches and techniques utilized by researchers for problem formulation and its solution are identified and tabulated in Table 3.

Table 3: Review of machine loading problems in FMS based on optimization techniques not falling in the above classification

			OTHER OPTIMIZATION TECHNIQUES		
1	1984	K. E. Stecke & T. L. Morin [67]	Single server closed queueing network model	➤ Balancing of workload	Maximizes expected production of FMS	Results are compared and contrasted with previous models of production systems
2	1986	K. E. Stecke [68]	Hierarchical approach	➤ Maximizing throughput	Nonlinear integer programs models	Ties with some previous results & use of the proposed models to solve realistic loading problems is discussed
3	1986	J. G. Shanthikumar & K. E. Stecke [69]	Dynamic approach	➤ Balancing of workload	Result maximizes expected production	results obtained here complement previous results from literature
4	1993	Y.-D. Kim [70]	Due-Date Based Loading methods	➤ Maximizing throughput	Results reported reduced tardiness and makespan & in-creased throughput	Computational tests
5	1997	H.-K. Roh & Y-D. Kim [71]	Due-Date Based Loading methods	➤ Minimizing total tardiness	Iterative approach performs better than others	Computational tests on randomly generated problems
6	1997	D. H. Lee, S. K. Lim, G. C. Lee, H. B. Jun, & Y. D. Kim [72]	Iterative algorithms	➤ Minimizing subcontracting costs	Solved part selection and loading problems	computational experiments on randomly generated test problems
7	1997	Y. D. Kim and C. A. Yano [73]	Queueing network model	➤ Maximizing throughput ➤ Maximizing make span ➤ Balancing of workload	Reducing number of machine groups and balancing workloads among machines help to reduce make span	Computational results are produced
8	1998	D.-H. Lee & Y.-D. Kim [74]	Iterative procedures	➤ Minimizing earliness ➤ Minimizing tardiness ➤ Minimizing subcontracting costs	Computational experiments on randomly generated test problems are produced	computational experiments are done on randomly generated test problems and the results are compared with existing results
9	1999	J. N. D. Gupta, L. H. S. Luong, & V. H. Nguyen [75]	Dispatching approach	➤ Minimizing make spans ➤ Minimizing average flow time ➤ Minimizing tardiness	Satisfactory performance of given dispatching algorithm	Simulation results are compared with existing results

| 10 | 2000 | S. Rahimifard & S.T. Newman [76] | Combined machine loading (CML) algorithms | ➤ Meeting delivery dates ➤ Minimising production costs | Adoption of algorithms within an application is dependent on number of manufacturing constraints | Computational results are produced and performance measure is carried out in virtual environment |
| 11 | 2012 | W. F. Mahmudy, R. M. Marian, & L. H. S. Luong [77] | Real coded genetic algorithms (RCGA) | ➤ Maximizing throughput ➤ Minimizing system unbalance | RCGA improves FMS performance & minimizes required computational time | Results are compared to the previous literature work |

The tick marks (√) shows the density of repetitive occurrence of the optimization techniques and approaches for solving the machine loading problem.

Abbreviations used in Table 5:

- Genetic Algorithm (GA): GA, Hybrid GA, Constraint based GA, Constraint-chromosome GA, Real coded GA, integrated approach based on GA

- Heuristic Algorithm (HA): HA, Fast HA, Fuzzy based HA, GA based HA, Hybrid TS and SA based HA, Lagrangian based HA, GA and PSO based Meta-hybrid HA, multi stage programming approach based HA

- Simulated annealing (SA): SA, Constraints-Based Fast SA, GA based SA, Hybrid GA-SA & SA-TS algorithm

Table 4: Objectives of machine loading in FMS

Sr.	Year	Researcher Name	1	2	3	4	5	6	7	8	9	10	11	12	13	14	15	16	17	18	19	20	21	22	23	24	25	26	27
1	1983	K. E. Stecke & F. B. Talbot [2]			√						√	√																	
2	1984	K. E. Stecke & T. L. Morin [67]			√																								
3	1985	K. Shankar & Y. J. Tzen [14]	√		√													√											
4	1986	M. Berrada & K. E. Stecke [26]			√																								
5	1986	K. E. Stecke [68]		√																									
6	1986	J.G.S. Kumar & K. E. Stecke [69]			√																								
7	1986	T. J. Greene & R. Sadowski [29]				√											√	√											
8	1987	S.C. Sarin & C.S. Chen [30]							√																				
9	1988	J. A. Ventura et al. [15]				√																							
10	1989	K. Shankar & A. Srinivasulu [27]	√							√											√								
11	1990	B. Ram et al. [16]						√																					
12	1990	K. M. Bretthauer et al. [31]													√														
13	1990	H. C. Co et al. [32]				√																							
14	1990	M. Liang & S. P. Dutt [33]							√																				
15	1992	Y. D. Kim & C. A. Yano [28]	√																										
16	1992	S. K. Mukhopadhyay et al. [17]	√	√																									
17	1993	K. Kato et al. [18]	√																							√			
18	1993	Ming Liang [34]	√																										
19	1993	Y-D. Kim [70]	√																										
20	1994	Ming Liang [35]	√					√																					
21	1995	E. K. Stecke & F. B. Talbot [19]		√						√	√																		
22	1997	M. K. Tiwari et al. [20]	√	√																									

#	Year	Author												
23	1997	V. N. Hsu & R. D. Matta [36]						√						
24	1997	H.-K. Roh & Y.-D. Kim [71]						√						
25	1997	D. H. Lee et al. [72]							√					
26	1997	Y. D. Kim and C. A. Yano [73]			√	√	√							
27	1998	S. K. Mukhopadhyay et al. [44]	√											
28	1998	D.-H. Lee & Y.-D. Kim [74]						√		√				√
29	1998	G. K. Nayak & D. Acharya [21]							√			√		
30	1998	T. J. Sawik [37]			√			√						
31	1999	F. Guerrero et al. [38]			√									
32	1999	J. N. D. Gupta et al. [75]				√		√		√				
33	2000	N. Kumar & K. Shanker [49]		√					√					
34	2000	D.-H. Lee & Y.-D. Kim [22]												√
35	2000	S. Rahimifard & S. Newman [76]					√	√						
36	2000	M. K. Tiwari & N. Vidyarthi [55]		√	√									
37	2001	N. Kumar & K. Shanker [39]			√									
38	2001	N. K. Vidyarthi & M. K. Tiwari [64]		√	√									
39	2002	H. Yong & Z. Wu [50]				√								
40	2003	M. Gamila & S. Motavalli [40]									√		√	
41	2004	R. R. Kumar et al. [65]		√	√									
42	2004	T. Sawik [41]									√			
43	2004	R. Swarnkar & M. K. Tiwari [45]		√	√									
44	2005	M. Aldaihani & M. Savsar [46]					√							
45	2006	N. Nagarjunaa et al. [23]		√										
46	2006	M. Goswami & M. Tiwari [24]		√	√									
47	2006	M. K. Tiwari et al. [47]		√	√									
48	2006	A. Kumar et al. [51]				√		√	√		√	√	√	
49	2007	A. Turkcan et al. [52]					√	√						
50	2007	M. K. Tiwari et al. [25]		√	√									
51	2007	S. Biswas & S. Mahapatra [61]		√										
52	2008	A. Prakash et al. [66]		√	√									
53	2008	S. Biswas & S. Mahapatra [62]		√										
54	2008	S. Ponnambalam & L. Kiat [63]		√	√									
55	2008	V. Tyagi & A. Jain [53]		√										
56	2009	M. Yogeswaran et al. [56]		√	√									
57	2010	S. K. Mandal et al. [57]		√	√	√								√
58	2011	M. I. Mgwatua [42]			√	√								
59	2012	V. M Kumar et al. [58]		√	√									
60	2012	C. Basnet [59]		√										
61	2012	M. Arıkan & S. Erol [48]		√		√				√				
62	2012	U. K. Yusof et al. [54]		√	√									
63	2012	D. Kosucuoglu & U. Bilge [60]											√	
64	2012	A. M. Abazari et al. [43]		√										
65	2012	W. F. Mahmudy et al. [77]		√	√									

- Mathematical programming (MP): MP, Linear MP, Non-linear MP, GA based MP
- Swarm Optimization (SO): SO, Particle SO (PSO), Modified PSO
- Queueing network model (QNM): QNM, Single server closed QNM
- Mixed-integer programming (MIP): MIP, GA based MIP
- Branch and bound algorithms (B&BA) : B&BA, New B&BA
- Integer programming (IP): IP, linear IP
- Non-linear programming
- Stochastic model

Table 5: Optimization techniques used for solving machine loading problems in FMS

Sr.	Year	Researcher Name	1	2	3	4	5	6	7	8	9	10	11	12	13	14	15	16	17	18	19	20	21
1	1983	K. E. Stecke & F. B. Talbot [2]	√																				
2	1984	K. E. Stecke & T. L. Morin [67]				√																	
3	1985	K. Shankar & Y. J. Tzen [14]	√																				
4	1986	M. Berrada & K. E. Stecke [26]							√														
5	1986	K. E. Stecke [68]														√							
6	1986	J.G.S. Kumar & K. E. Stecke [69]																					√
7	1986	T. J. Greene & R. Sadowski [29]						√															
8	1987	S. C. Sarin & C. S. Chen [30]			√																		
9	1988	J. A. Ventura et al. [15]	√																				
10	1989	K. Shankar & A. Srinivasulu [27]	√														√						
11	1990	B. Ram et al. [16]	√																				
12	1990	K. M. Bretthauer et al. [31]									√												
13	1990	H. C. Co et al. [32]						√															
14	1990	M. Liang & S. P. Dutt [33]						√															
15	1992	Y. D. Kim & C. A. Yano [28]								√													
16	1992	S. K. Mukhopadhyay et al. [17]	√																				
17	1993	K. Kato et al. [18]	√																				
18	1993	Ming Liang [34]										√											
19	1993	Y.-D. Kim [70]																			√		
20	1994	Ming Liang [35]										√											
21	1995	E. K. Steeke & F. B. Talbot [19]	√																				
22	1997	M. K. Tiwari et al. [20]	√																				
23	1997	V. N. Hsu & R. D. Matta [36]	√																				

No	Year	Author											
24	1997	H.-K. Roh & Y.-D. Kim [71]											√
25	1997	D. H. Lee et al. [72]									√		
26	1997	Y. D. Kim and C. A. Yano [73]						√					
27	1998	S. K. Mukhopadhyay et al. [44]			√								
28	1998	D.-H. Lee & Y.-D. Kim [74]									√		
29	1998	G. K. Nayak & D. Acharya [21]		√		√							
30	1998	T. J. Sawik [37]							√		√		
31	1999	F. Guerrero et al. [38]						√					
32	1999	J. N. D. Gupta et al. [75]										√	
33	2000	N. Kumar & K. Shanker [49]		√									
34	2000	D.-H. Lee & Y.-D. Kim [22]		√									
35	2000	S. Rahimifard & S. Newman [76]										√	
36	2000	M. K. Tiwari & N. Vidyarthi [55]		√									
37	2001	N. Kumar & K. Shanker [39]						√					
38	2001	N. K. Vidyarthi & M. K. Tiwari [64]		√									
39	2002	H. Yong & Z. Wu [50]	√					√					
40	2003	M. Gamila & S. Motavalli [40]											
41	2004	R. R. Kumar et al. [65]											√
42	2004	T. Sawik [41]						√					
43	2004	R. Swarnkar & M. K. Tiwari [45]		√									
44	2005	M. Aldaihani & M. Savsar [46]								√			
45	2006	N. Nagarjunaa et al. [23]		√									
46	2006	M. Goswami & M. Tiwari [24]		√									
47	2006	M. K. Tiwari et al. [47]			√								
48	2006	A. Kumar et al. [51]	√										
49	2007	A. Turkcan et al. [52]	√										
50	2007	M. K. Tiwari et al. [25]		√									
51	2007	S. Biswas & S. Mahapatra [61]					√						
52	2008	A. Prakash et al. [66]								√			
53	2008	S. Biswas & S. Mahapatra [62]					√						
54	2008	S. Ponnambalam & L. Kiat [63]					√						
55	2008	V. Tyagi & A. Jain [53]	√										
56	2009	M. Yogeswaran et al. [56]			√								
57	2010	S. K. Mandal et al. [57]	√	√	√								
58	2011	M. I. Mgwatua [42]					√						
59	2012	V. M Kumar et al. [58]			√		√						
60	2012	C. Basnet [59]	√										
61	2012	M. Arıkan & S. Erol [48]			√								
62	2012	U. K. Yusof et al. [54]	√										
63	2012	D. Kosucuoglu & U. Bilge [60]						√					
64	2012	A. M. Abazari et al. [43]			√								
65	2012	W. F. Mahmudy et al. [77]	√										

- Modified immune algorithm
- Approximative lexicographic approach

- Iterative algorithms
- Hierarchical approach
- Branch & backtrack procedure
- Combined machine loading algorithms
- Dispatching approach
- Due-Date Based Loading methods
- Dynamic approach
- Fuzzy Logic

CONCLUSION ARRIVED ON MACHINE LOADING OBJECTIVES AND OPTIMIZATION TECHNIQUES IN FMS

Detailed study of the machine loading problem is conducted by the authors. The conclusions of the research throttled are divided into three sections as below.

Conclusion on Machine Loading Objectives

On exhaustive study, twenty eight loading objectives are observed in the reviewed literature. Tick marks ($\sqrt{}$) in Table 4 are showing the density for repeatability of the machine loading objectives, which concludes that a research with maximum loading objectives is still required for solving the machine problem. Maximizing expected production rate (throughput) & unbalancing the workload per machine for a system of groups of pooled machines of unequal sizes are the two objectives on which most of the researchers have worked. Balancing of workload on machines for a system of groups of pooled machines of equal sizes is the second most researched loading objective. Minimizing make span is the third most researched loading objective. Minimizing job tardiness is fourth loading objective in the order.

Minimizing mean job flow time & minimizing production cost are found at fifth position in the order. Loading objectives observed at sixth rank are maximizing profitability, maximizing the assigned workload, maximizing the part types in each batch, maximizing utilization of system, minimizing subcontracting costs and minimizing the total number of cutting tools required. Material handling time, maximizing routing flexibility of the batches, minimizing earliness, minimizing mean lateness, minimizing mean machine idle time, minimizing overall machining cost, minimizing production time, minimizing the effect of breakdowns, minimizing the maximum workload of the machines, minimizing the number of late jobs, minimizing total flexible manufacturing cell cost per unit of production, minimizing total inter-station transfer time, minimizing total processing time, minimizing part movements and minimisation of the total distance travelled by parts during their production are the loading objectives that are least considered.

Conclusion on Optimization Techniques in FMS

The categorized literature review concludes that the researcher's major emphasis and contribution are towards the use and application of global optimization techniques and with natural optimization techniques, too. Heuristic Algorithms is the mostly used optimization technique by researchers, followed by Genetic Algorithms (GA). Mixed Integer Programming (MIP) & Simulated Annealing (SA) approach are the third mostly used optimization techniques. Linear Mathematical Programming (LMP) is next in the queue succeeded by Integer Programming (IP). At sixth level is Particle Swarm Optimization (PSO) approach. The least used optimization techniques are Tabu search, Swarm Optimization Approach, Branch and backtrack procedure, Branch and bound approach, Combined machine loading (CML) algorithms, Dispatching approach, Due-Date Based methods, Dynamic approach, Fuzzy Logic, Global criterion approach, Hierarchical approach, Artificial immune algorithm, Iterative algorithms, Lexicographic approach, Non-linear programming, Queueing network model and Stochastic model.

Conclusion on validation approaches

A few research problems are solved and the results are compared with previous research results. The results are validated by comparing with literature available results.

Methodologies Findings and Interpretations

A problem when solved for a limited or less number of objectives, it is rather a customized solution for a problem. For general solution, the problem needs to be solved for all possible objectives. On extreme analysis of the machine loading problem and objectives, and on discussion with the academicians and industrialists, the authors emphasise to solve the loading problem for maximization of throughput, part types in a batch, routing flexibility, balancing/ unbalancing of system and workload, and minimization of make-span, delivery dates (covering lateness, tardiness and earliness), part movements, subcontracting costs, machine processing time, tool magazine capacity, number of cutting tools required, breakdowns, non-splitting of jobs, time spend by job on machines in one study. Machine loading problem should be solved for general solution to the problem, for maximum number of objectives. All these objectives are having a common goal of optimizing the production and manufacturing costs.

The literature review reports the application of heuristics, global optimization techniques and some other optimization techniques for solving the loading problem for the listed objectives. Among these approaches, the global optimization techniques were more frequently adopted and the results as founded by the researchers were more accurate and acceptable. Based on regressive analysis of the available literature, and skills and concluding remarks, the authors suggest for the use of natural optimization techniques like swarm optimization for further research. The results of swarm optimization were found more reliable and acceptable as compared to GA, and PSO has attractive characteristics. PSO retains knowledge of all previous particles, which is destroyed in GA when the population changes. PSO is a mechanism of constructive cooperation and information-sharing between particles. Due to the simple concept, ease of implementation, and quick convergence, PSO has gained much attention and has been successfully applied to a wide range of applications.

RESEARCH GAPS AND SCOPE OF RESEARCH IN LOADING OF MACHINES IN FMS

There exists a research gap among the literature available. There are several future scopes that are still not worked out, or still to be worked in a more optimized manner. Based on our observation and exhaustive study such revealed research gap are listed below: Need of integration of loading with other decisions in the neighbourhood of loading (K. Shankar & A. K. Agrawal, 1991); need to reduce excessive computing times (Y. D. Kim & C. A. Yano, 1989); further need of optimization (N. K. Vidyarthi & M. K. Tiwari, 2001, M. K. Tiwari et al., 2007; Amir Musa Abazari et al., 2012); research is required to develop planning softwares (D. H. Lee et al., 1997); PLC controller needs to be enhanced (M. C. Zhou et al., 1993); waiting time for parts and idling time for machines need attention [Mussa I. Mgwatu, 2011]; research by imposing constraints on the availability of resources i.e. jigs, fixtures, pallets, material handling devices needs to be carried out (K. Kato, 1993, N. K. Vidyarthi & M. K. Tiwari, 2001; N. Nagarjuna et al., 2006; Akhilesh Kumar et al., 2006; M. K. Tiwari et al., 2007; Sandhyarani Biswas & S. S. Mahapatra, 2007; Sandhyarani Biswas & S. S. Mahapatra, 2008; Santosh Kumar Mandal et al. 2010; Amir Musa Abazari et al., 2012); new solution methodology needs to be proposed (Santosh Kumar Mandal et al. 2010); need of AI in the field of FMS

) Chinyao Low et al., 2006; Sandhyarani Biswas & S. S. Mahapatra, 2008); Need to use dedicated robot (Majid M. Aldaihani & Mehmet Savsar, 2005); need of simulation studies for FMS (K. Shankar & A. K. Agrawal, 1991; N. K. Vidyarthi & M. K. Tiwari, 2001). Availability of a number of research gaps and that too identified by various eminent researchers from time to time evacuates the need of vast research for solving the observed PPC problems i.e. machine loading problems in FMS.

The authors are working to solve the loading problem with more number of objectives in a single study and for the development of knowledge base system for the machine loading problem. The authors suggest for the development of a knowledge base for all five productions planning problems; part type selection problem, machine grouping problem, production ratio problem, resource allocation problem and loading problem in a single study incorporating the individual objectives of the five individual problems and their respective technological and capacity constraints.

REFERENCES

1. Stecke, K.E. (1983) Formulation and Solution of Nonlinear Integer Production Planning Problems for Flexible Manufacturing Systems. Management Science, 29, 273-288. http://dx.doi.org/10.1287/mnsc.29.3.273

2. Stecke, K.E. and Talbot, F.B. (1983) Heuristic Loading Algorithms for Flexible Manufacturing Systems. Proceedings of the Seventh International Conference on Production Research, Windsor, 22-24 August 1983.

3. Muhlenbein, H. (1992) Parallel Genetic Algorithms in Combinatorial Optimization. In: Balci, O., Sharda, R. and Zenios, S.A., Eds., Computer Science and Operations Research: New Developments in Their Interfaces, Pergamon Press, Oxford, 441-456.

4. Russell, S.J. and Norvig, P. (2002) Artificial Intelligence: A Modern Approach. Second Edition, Prentice Hall, Englewood Cliffs.

5. Rastrigin, L.A. (1963) The Convergence of the Random Search Method in the External Control of Many-Parameter System. Automation and Remote Control, 24, 1337-1342.

6. Metropolis, N., Rosenbluth, A.W., Rosenbluth, M.N., Teller, A.H. and Teller, E. (1953) Equation of State Calculations by Fast Computing Machines. The Journal of Chemical Physics, 21, 1087-1092. http://dx.doi.org/10.1063/1.1699114

7. Glover, F. (1989) Tabu Search—Part I. Operations Research Society of America (ORSA). Journal on Computing, 1, 90-206. http://dx.doi.

org/10.1287/ijoc.1.3.190

8. Moscato, P. (1989) On Evolution, Search, Optimization, Genetic Algorithms and Martial Arts: Towards Memetic Algorithms. Technical Report C3P 826, Caltech Con-Current Computation Program 158-79, California Institute of Technology, Pasadena.

9. Holland, J.H. (1962) Outline for a Logical Theory of Adaptive Systems. Journal of the ACM, 9, 297-314. http://dx.doi.org/10.1145/321127.321128

10. Holland, J.H. and Reitman, J.S. (1977) Cognitive Systems Based on Adaptive Algorithms. ACM SIGART Bulletin, 63, 49. http://dx.doi.org/10.1145/1045343.1045373

11. Friedberg, R.M. (1958) A Learning Machine: Part I. IBM Journal of Research and Development, 2, 2-13. http://dx.doi.org/10.1147/rd.21.0002

12. Dorigo, M., Maniezzo, V. and Colorni, A. (1996) The Ant System: Optimization by a Colony of Cooperating Agents. IEEE Transactions on Systems, Man, and Cybernetics Part B: Cybernetics, 26, 29-41. http://dx.doi.org/10.1109/3477.484436

13. Eberhart, R.C. and Kennedy, J. (1995) A New Optimizer Using Particle Swarm Theory. Proceedings of the Sixth International Symposium on Micro Machine and Human Science, Nagoya, 4-6 October 1995, 39-43. http://dx.doi.org/10.1109/MHS.1995.494215

14. Shankar, K. and Tzen, Y.J.J. (1985) A Loading and Dispatching Problem in a Random Flexible Manufacturing System. International Journal of Production Research, 23, 579-595. http://dx.doi.org/10.1080/00207548508904730

15. Ventura, J.A., Chen, F.F. and Leonard, M.S. (1988) Loading Tools to Machines in Flexible Manufacturing Systems. Computers & Industrial Engineering, 15, 223-230.

16. Ram, B., Sarin, S. and Chen, C.S. (1990) A Model and Solution Approach for the Machine Loading and Tool Allocation Problem in FMS. International Journal of Production Research, 28, 637-645.

17. Mukhopadhyay, S.K., Midha, S. and Murlikrishna, V. (1992) A Heuristic Procedure for Loading Problem in Flexible Manufacturing Systems. International Journal of Production Research, 30, 2213-2228. http://dx.doi.org/10.1080/00207549208948146

18. Kato, K., Oba, F. and Hashimoto, F. (1993) Loading and Batch Formation in Flexible Manufacturing Systems. Control Engineering Practice, 1, 845-850. http://dx.doi.org/10.1016/0967-0661(93)90252-M

19. Steeke, E.K. and Talbot, F.B. (1995) Heuristics for Loading Flexible

Manufacturing Systems, Flexible Manufacturing Systems: Recent Developments. Elsevier Science B.V., Amsterdam, 171-176.

20. Tiwari, M.K., Hazarika, B., Vidyarthi, N.K., Jaggi, P. and Mukhopadhyay, S.K. (1997) A Heuristic Solution Approach to the Machine Loading Problem of FMS and Its Petri Net Model. International Journal of Production Research, 35, 2269-2284. http://dx.doi.org/10.1080/002075497194840

21. Nayak, G.K. and Acharya, A.D. (1998) Part Type Selection, Machine Loading and Part Type Volume Determination in FMS Planning. International Journal of Production Research, 36, 1801-1824. http://dx.doi.org/10.1080/002075498192977

22. Lee, D.H. and Kim, Y.-D. (2000) Loading Algorithms for Flexible Manufacturing Systems with Partially Grouped Machines. IIE Transactions, 32, 33-47.

23. Nagarjuna, N., Mahesh, O. and Rajagopal, K. (2006) A Heuristic Based on Multi-Stage Programming Approach for Machine-Loading Problem in a Flexible Manufacturing System. Robotics and Computer-Integrated Manufacturing, 22, 342-352. http://dx.doi.org/10.1016/j.rcim.2005.07.006

24. Goswami, M. and Tiwari, M.K. (2006) A Reallocation-Based Heuristic to Solve a Machine Loading Problem with Material Handling Constraint in a Flexible Manufacturing System. International Journal of Production Research, 44, 569-588.

25. Tiwari, M.K., Saha, J. and Mukhopadhyay, S.K. (2007) Heuristic Solution Approaches for Combined-Job Sequencing and Machine Loading Problem in Flexible Manufacturing Systems. International Journal of Advanced Manufacturing Technology, 31, 716-730.

26. Berrada, M. and Stecke, K.E. (1986) A Branch and Bound Approach for Machine Load Balancing in Flexible Manufacturing Systems. Management Science, 32, 1316-1335. http://dx.doi.org/10.1287/mnsc.32.10.1316

27. Shankar, K. and Srinivasulu, A. (1989) Some Selection Methodologies for Loading Problems in a Flexible Manufacturing System. International Journal of Production Research, 27, 1019-1034. http://dx.doi.org/10.1080/00207548908942605

28. Kim, Y.-D. and Yano, C.A. (1994) A New Branch and Bound Algorithm for Loading Problems in Flexible Manufacturing Systems. International Journal of Flexible Manufacturing Systems, 6, 361-381. http://dx.doi.org/10.1007/BF01324801

29. Greene, T.J. and Sadowski, R.P. (1986) A Mixed Integer Programming

for Loading and Scheduling Multiple Manufacturing Cells. European Journal of Operation Research, 24, 379-386.

30. http://dx.doi.org/10.1016/0377-2217(86)90031-7

31. Sarin, S.C. and Chen, C.S. (1987) The Machine Loading and Tool Allocation Problem in a Flexible Manufacturing System. International Journal of Production Research, 25, 1081-1094. http://dx.doi.org/10.1080/00207548708919897

32. Bretthauer, K.M. and Venkataramanan, M.A. (1990) Machine Loading and Alternate Routing in a Flexible Manufacturing System. Computers and Industrial Engineering, 18, 341-350. http://dx.doi.org/10.1016/0360-8352(90)90056-R

33. Co, H.C., Biermann, J.S. and Chen, S.K. (1990) A Methodical Approach to the Flexible Manufacturing System Batching, Loading and Tool Configuration Problems. International Journal of Production Research, 28, 2171-2186. http://dx.doi.org/10.1080/00207549008942860

34. Liang, M. and Dutt, S.P. (1990) A Mixed-Integer Programming Approach to the Machine Loading and Process Planning Problem in a Process Layout Environment. International Journal of Production Research, 28, 1471-1484. http://dx.doi.org/10.1080/00207549008942806

35. Liang, M. (1993) Part Selection, Machine Loading and Machining Speed Selection in Flexible Manufacturing Systems. Computers and Industrial Engineering, 25, 259-262. http://dx.doi.org/10.1016/0360-8352(93)90270-8

36. Liang, M. (1994) Integrating Machining Speed, Part Selection and Machine Loading Decisions in Flexible Manufacturing Systems. Computers & Industrial Engineering, 26, 599-608.

37. Hsu, V.N. and De Matta, R. (1997) An Efficient Heuristic Approach to Recognize the Infeasibility of a Loading Problem. International Journal of Manufacturing Systems, 9, 31-50.

38. awik, T.J. (1998) A Lexicographic Approach to Bi-Objective Loading of a Flexible Assembly System. European Journal of Operational Research, 107, 656-668. http://dx.doi.org/10.1016/S0377-2217(97)00091-X

39. Guerreore, F., Lozano, S., Koltai, T. and Larraneta, J. (1999) Machine Loading and Part Type Selection in Flexible Manufacturing System. International Journal of Production Research, 37, 1303-1317. http://dx.doi.org/10.1080/002075499191265

40. Kumar, N. and Shanker, K. (2001) Comparing the Effectiveness of Workload Balancing Objectives in FMS Loading. International Journal

of Production Research, 39, 843-871.

41. Gamila, M.A. and Motavalli, S. (2003) A Modeling Technique for Loading and Scheduling Problems in FMS. Robotics and Computer Integrated Manufacturing, 19, 45-54.

42. Sawik, T. (2004) Loading and Scheduling of a Flexible Assembly System by Mixed Integer Programming. European Journal of Operational Research, 154, 1-19. http://dx.doi.org/10.1016/S0377-2217(02)00795-6

43. Mgwatua, M.I. (2011) Interactive Decisions of Part Selection, Machine Loading, Machining Optimisation and Part Scheduling Sub-Problems for Flexible Manufacturing Systems. International Transaction Journal of Engineering, Management, & Applied Sciences & Technologies, 2, 93-109.

44. Abazari, A.M., Solimanpur, M. and Sattari, H. (2012) Optimum Loading of Machines in a Flexible Manufacturing System Using a Mixed-Integer Linear Mathematical Programming Model and Genetic Algorithm. Computers & Industrial Engineering, 62, 469-478. http://dx.doi.org/10.1016/j.cie.2011.10.013

45. Mukhopadhyay, S.K., Singh, M.K. and Srivastava, R. (1998) FMS Loading: A Simulated Annealing Approach. International Journal of Production Research, 36, 1529-1547. http://dx.doi.org/10.1080/002075498193156

46. Swarnkar, R. and Tiwari, M.K. (2004) Modeling Machine Loading Problem of FMSs and Its Solution Methodology Using a Hybrid Tabu Search and Simulated Annealing-Based Heuristic Approach. Robotics and Computer-Integrated Manufacturing, 20, 199-209. http://dx.doi.org/10.1016/j.rcim.2003.09.001

47. Aldaihani, M.M. and Savsar, M. (2005) A Stochastic Model for the Analysis of a Two-Machine Flexible Manufacturing Cell. Computers & Industrial Engineering, 49, 600-610. http://dx.doi.org/10.1016/j.cie.2005.09.002

48. Tiwari, M.K., Kumar, S., Kumar, S., Prakash and Shankar, R. (2006) Solving Part-Type Selection and Operation Allocation Problems in an FMS: An Approach Using Constraints-Based Fast Simulated Annealing Algorithm. IEEE Transactions on Systems, Man, and Cybernetics—Part A: Systems and Humans, 36, 1170-1184.

49. Arikan, M. and Erol, S. (2012) A Hybrid Simulated Annealing-Tabu Search Algorithm for the Part Selection and Machine Loading Problems in Flexible Manufacturing Systems. International Journal of Advanced Manufacturing Technology, 59, 669-679. http://dx.doi.org/10.1007/s00170-011-3506-0

50. Kumar, N. and Shanker, K. (2000) A Genetic Algorithm for FMS Part Type Selection and Machine Loading. International Journal of Production Research, 38, 3861-3887.

51. Yong, H.H. and Wu, Z.M. (2002) GA-Based Integrated Approach to FMS Part Type Selection and Machine Loading Problem. International Journal of Production Research, 40, 4093-4110. http://dx.doi.org/10.1080/00207540210146972

52. Kumar, A., Prakash, Tiwari, M.K., Shankar, R. and Baveja, A. (2006) Solving Machine-Loading Problem of a Flexible Manufacturing System with Constraint-Based Genetic Algorithm. European Journal of Operational Research, 175, 1043-1069. http://dx.doi.org/10.1016/j.ejor.2005.06.025

53. Turkcan, A., Akturk, M.S. and Storer, R.H. (2007) Due Date and Costbased FMS Loading, Scheduling and Tool Management. International Journal of Production Research, 45, 1183-1213.

54. Tyagi, V. and Jain, A. (2008) Assessing the Effectiveness of Flexible Process Plans for Loading and Part Type Selection in FMS. Advances in Production Engineering & Management, 3, 27-44.

55. Yusof, U.K., Budiarto, R. and Deris, S. (2012) Constraint-Chromosome Genetic Algorithm for Flexible Manufacturing System Machine-Loading Problem. International Journal of Innovative Computing, Information and Control, 8, 1591-1609.

56. Tiwari, M.K. and Vidyarthi, N.K. (2000) Solving Machine Loading Problem in Flexible Manufacturing System Using Genetic Algorithm Based Heuristic Approach. International Journal of Production Research, 38, 3357-3384. http://dx.doi.org/10.1080/002075400418298

57. Yogeswaran, M., Ponnambalam, S.G. and Tiwari, M.K. (2009) An Efficient Hybrid Evolutionary Heuristic Using Genetic Algorithm and Simulated Annealing Algorithm to Solve Machine Loading Problem in FMS. International Journal of Production Research, 47, 5421-5448.

58. Mandal, S.K., Pandey, M.K. and Tiwari, M.K. (2010) Incorporating Dynamism in Traditional Machine Loading Problem: An AI-Based Optimization Approach. International Journal of Production Research, 48, 3535-3559. http://dx.doi.org/10.1080/00207540902814306

59. Kumar, V.M., Murthy, A.N.N. and Chandrashekar, K. (2012) A Hybrid Algorithm Optimization Approach for Machine Loading Problem in Flexible Manufacturing System. Journal of Industrial Engineering International, 8, 3. http://dx.doi.org/10.1186/2251-712X-8-3

60. Basnet, C. (2012) A Hybrid Genetic Algorithm for a Loading Problem in Flexible Manufacturing Systems. International Journal of Production Research, 50, 707-718.

61. Kosucuoglu, D. and Bilge, U. (2012) Material Handling Considerations in the FMS Loading Problem with Full Routing Flexibility. International Journal of Production Research, 50, 6530-6552.

62. Biswas, S. and Mahapatra, S.S. (2007) Machine Loading in Flexible Manufacturing System: A Swarm Optimization Approach. Proceedings of the Eighth International Conference on Operations and Quantitative Management, Bangkok, 17-20 October 2007.

63. Biswas, S. and Mahapatra, S.S. (2008) Modified Particle Swarm Optimization for Solving Machine Loading Problems in Flexible Manufacturing Systems. International Journal of Advanced Manufacturing Technology, 39, 931-942.

64. Ponnambalam, S.G. and Kiat, L.S. (2008) Solving Machine Loading Problem in Flexible Manufacturing Systems Using Particle Swarm Optimization. World Academy of Science, Engineering and Technology, 39, 14-19.

65. Vidyarthi, N.K. and Tiwari, M.K. (2001) Machine Loading Problem of FMS: A Fuzzy-Based Heuristic Approach. International Journal of Production Research, 39, 953-979. http://dx.doi.org/10.1080/00207540010010244

66. Kumar, R.R., Singh, A.K. and Tiwari, M.K. (2004) A Fuzzy Based Algorithm to Solve the Machine-Loading Problems of a FMS and Its Neuro Fuzzy Petri Net Model. International Journal of Advanced Manufacturing Technology, 23, 318-341. http://dx.doi.org/10.1007/s00170-002-1499-4

67. Prakash, A., Khilwani, N., Tiwari, M.K. and Cohen, Y. (2008) Modified Immune Algorithm for Job Selection and Operation Allocation Problem in Flexible Manufacturing Systems. Advances in Engineering Software, 39, 219-232. http://dx.doi.org/10.1016/j.advengsoft.2007.01.024

68. Stecke, K.E. and Morin, T.L. (1985) The Optimality of Balancing Workloads in Certain Types of Flexible Manufacturing Systems. European Journal of Operational Research, 20, 68-82.

69. Stecke, K.E. (1986) A Hierarchical Approach to Solving Grouping and Loading Problems of Flexible Manufacturing Systems. European Journal of Operational Research, 24, 369-378. http://dx.doi.org/10.1016/0377-2217(86)90030-5

70. Shanthikumar, J.G. and Stecke, K.E. (1986) Reducing Work in Progress Inventory in Certain Classes of Flexible Manufacturing Systems.

European Journal of Operation Research, 26, 266-271. http://dx.doi. org/10.1016/0377-2217(86)90189-X

71.   Kim, Y.-D. (1993) A Study on Surrogate Objectives for Loading a Certain Type of Flexible Manufacturing Systems. International Journal of Production Research, 31, 381-392. http://dx.doi.org/10.1016/0377-2217(86)90189-X

72.   Roh, H.-K. and Kim, Y.-D. (1997) Due-Date Based Loading and Scheduling Methods for a Flexible Manufacturing System with an Automatic Tool Transporter. International Journal of Production Research, 35, 2989-3004.

73.   Lee, D.-H., Lira, S.-K., Lee, G.-C., Jun, H.-B. and Kim, Y.-D. (1997) Multi-Period Part Selection and Loading Problems in Flexible Manufacturing Systems. Computers & Industrial Engineering, 33, 541-544.

74.   Kim, Y.D. and Yano, C.A. (1997) Impact of Throughput Based Objective and Machine Grouping Decisions on the Short-Term Performance of Flexible Manufacturing System. International Journal of Production Research, 35, 3303-3322. http://dx.doi.org/10.1080/002075497194084

75.   Lee, D.-H. and Kim, Y.-D. (1998) Iterative Procedures for Multi-Period Order Selection and Loading Problems in Flexible Manufacturing Systems. International Journal of Production Research, 36, 2653-2668. http://dx.doi.org/10.1080/002075498192418

76.   Gupta, J.N.D. (1999) Part Dispatching and Machine Loading in Flexible Manufacturing System Using Central Queues. International Journal of Production Research, 37, 1427-1435. http://dx.doi. org/10.1080/002075499191337

77.   Rahimifard, S. and Newman, S.T. (2000) Machine Loading Algorithms for the Elimination of Tardy Jobs in Flexible Batch Machining Applications. Journal of Materials Processing Technology, 107, 450-458.

78.   Mahmudy, W.F., Marian, R.M. and Luong, L.H.S. (2012) Solving Part Type Selection and Loading Problem in Flexible Manufacturing System Using Real Coded Genetic Algorithms—Part II: Optimization. World Academy of Science, Engineering and Technology, 69, 778-782.

Chapter 2

LEAD-FREE PIEZOELECTRIC DIAPHRAGM BIOSENSORS BASED ON MICRO-MACHINING TECHNOLOGY AND CHEMICAL SOLUTION DEPOSITION

Xiaomeng Li[1], Xiaoqing Wu[1], Peng Shi[1], and Zuo-Guang Ye[1, 2]

[1]Electronic Materials Research Laboratory, Key Laboratory of the Ministry of Education & International Center for Dielectric Research, Xi'an Jiaotong University, Xi'an 710049, China

[2]Department of Chemistry and 4D LABS, Simon Fraser University, Burnaby, BC V5A 1S6, Canada

ABSTRACT

In this paper, we present a new approach to the fabrication of integrated silicon-based piezoelectric diaphragm-type biosensors by using sodium potassium niobate-silver niobate (0.82KNN-0.18AN) composite lead-free thin film as the piezoelectric layer. The piezoelectric diaphragms were designed and fabricated by micro-machining technology and chemical solution deposition. The fabricated device was very sensitive to the mass changes caused by various targets attached on the surface of diaphragm. The measured mass sensitivity value was about 931 Hz/μg. Its good performance shows that the piezoelectric diaphragm biosensor can be used as a cost-effective platform for nucleic acid testing.

INTRODUCTION

Over the past few decades, researchers have focused on developing sensors with high sensitivity and low limits of detection. As a result, a wide range of biosensors with different operational modalities have been demonstrated. In particular, label-free biosensors are suitable for rapid and real-time sensing applications. Many types of label-free biosensors, such as nanogap and microneedle biosensors, have been widely studied [1,2,3,4,5,6,7,8]. In recent years, piezoelectric devices have been widely used as micro-pumps, actuators, transducers and biosensors [9,10,11,12,13,14,15,16,17,18]. The high sensitivity quartz crystal micro-balance (QCM) has been successfully used in biosensors for many years [19,20,21].

However, the conventional crystal or bulk ceramic technology is insufficient to meet the increasing requirements of miniaturization and integration. Recently, the development of micro-machined piezoelectric biosensors has become a focus of attention because of their merits such as compact size, high sensitivity, label-free operation, rapid response and compatibility with integrated circuit techniques [22,23,24,25,26,27]. Piezoelectric biosensors are mass-sensitive which is detected by measuring the shift of resonant frequency of the devices [25,28]. When the reaction between the recognition layer immobilized on the diaphragm and the captured target biological species happens, the total mass of the piezoelectric film area will change which results in a corresponding resonant frequency shift [29].

Nowadays, most piezoelectric devices use lead-based materials such as $Pb(Zr, Ti)O_3$ (PZT) and $Pb(Mg_{1/3}Nb_{2/3})O_3$-$PbTiO_3$(PMN-PT) because of their excellent piezoelectric properties. However, the toxicity of lead has raised concerns and research on lead-free piezoelectric biosensors is in great demanded. The traditional lead-free piezoelectric materials, such as $(K,Na)NbO_3$ (KNN) and $(Bi,Na)TiO_3$ (BNT), are not widely applied because of their weak electrical properties. During this work, we have successfully prepared a new lead-free piezoelectric biosensor by using the lead-free film of $0.82K_{0.5}Na_{0.5}NbO_3$-$0.18AgNbO_3$ (0.82KNN-0.18AN), which exhibited much improved electrical properties compared to pure KNN film [30].

RESULTS AND DISCUSSION

Characterization of the 0.82KNN-0.18AN Piezoelectric Layer

The crystallization and electrical properties of the 0.82KNN-0.18AN piezoelectric film were analysed. Figure 1 shows the XRD pattern of the 0.82KNN-0.18AN film, indicating that the film possessed a pure perovskite phase, and no obviously secondary phases were observed. It can be seen that the film exhibited (100) preferential orientation.

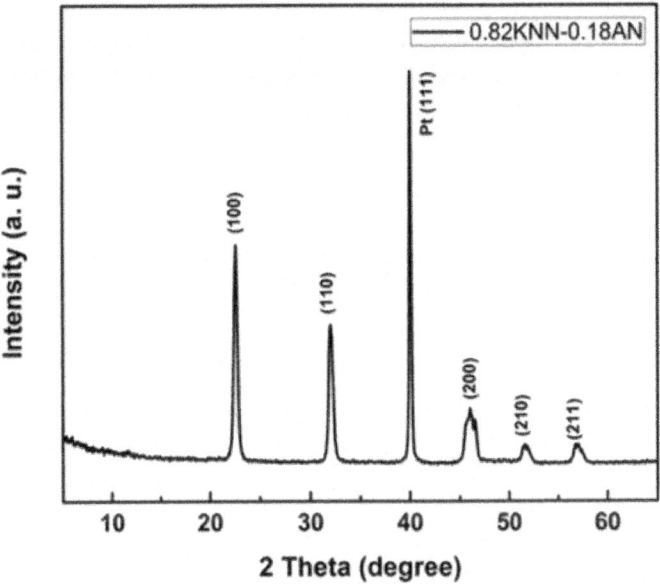

Figure 1: XRD pattern of the 0.82KNN-0.18AN film.

The surface and cross-sectional morphologies for the 0.82KNN-0.18AN film are shown in Figure 2. From the surface image, the average grain size of the 0.82KNN-0.18AN film was estimated to be about 200 nm. From the cross-sectional image, the 0.82KNN-0.18AN film thickness was calculated to be about 2.16 μm. The crack-free film showed a good structure which ensured the good quality of the biosensor.

Figure 2: SEM images of the (a) surface; and (b) cross-sectional morphologies for the 0.82KNN-0.18AN film.

The 0.82KNN-0.18AN film showed well defined P-E hysteresis loops while the pure KNN film deposited with the same method showed poorly shaped P-E hysteresis loops, as demonstrated in Figure 3. The remnant polarization (Pr) of the 0.82KNN-0.18AN film was calculated as 3.59 $\mu C/cm^2$, at the applied electric field of 550 kV/cm.

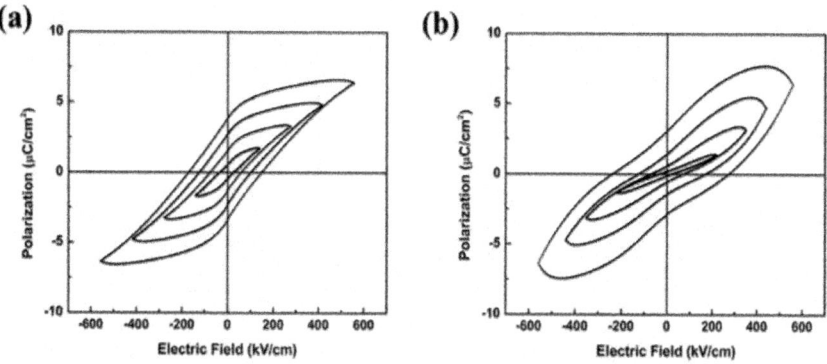

Figure 3: Hysteresis loops for the (a) 0.82KNN-0.18AN film; and (b) KNN film.

Characterization of the Piezoelectric Diaphragm Biosensor

Using the micro-machining process and deposition processing, piezoelectric diaphragm biosensors were successfully fabricated. As shown in Figure 4a, we can see that twenty individual biosensors were fabricated on a quarter of the 4" wafer. The SEM top and bottom views of one individual biosensor were shown in Figure 4b,c.

The structures of the diaphragm area, top electrode, connecting wire and connecting pad are shown in detail. The length of the biosensor was 1 mm and the length of the top electrode was 0.7 mm, which were the optimized parameters by the results of finite element simulation. The opening area of the chamber was much larger than that of the bottom diaphragm due to the anisotropic wet etching by KOH. The dipping and washing in the immobilization process were much more convenient due to this enlarged opening chamber. The cross section of the biosensor was shown in Figure 4d. Each layer of the multilayer structure was identified. The dense and crack-free 0.82KNN-0.18AN film ensured the high yield of the biosensor.

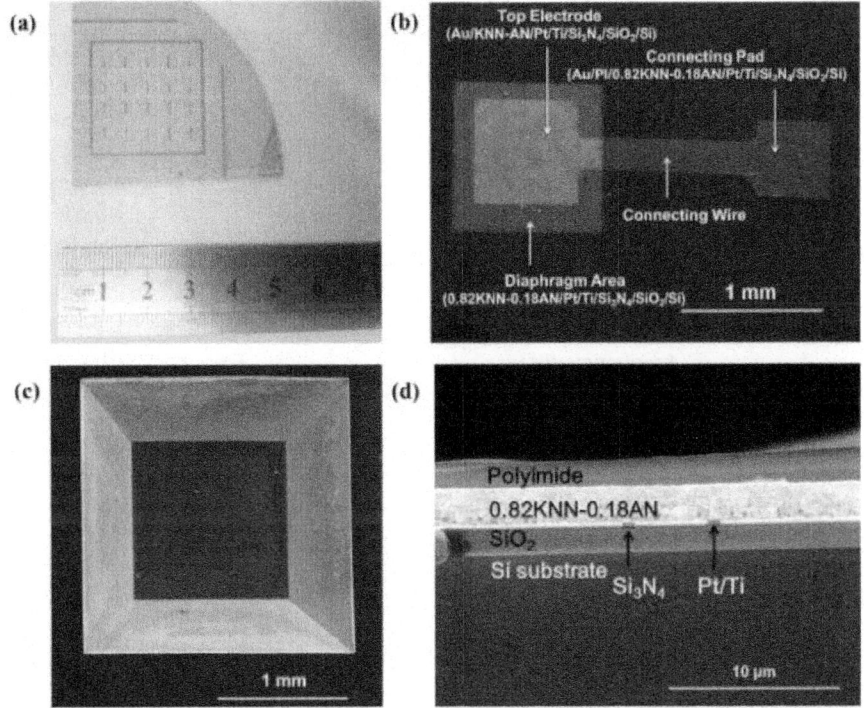

Figure 4: (a) Optical image; SEM images of the (b) front view; (c) back view; and (d) cross section of the fabricated biosensor.

The test of mass sensitivity was carried out using a biosensor with a residual silicon thickness of about 25 μm. The detailed steps were as follows: firstly, we cleaned and modified the surface of a Au top electrode with piranha solution (70% H_2SO_4: 30% H_2O_2). Secondly, we added 0.1 μL of single-stranded DNA (ssDNA) (5′-6-FAM/CACAACAGACGGGCACACACTACT/ C6-SH-3′) solution with a concentration of 1 μg/μL on the surface of top electrode via Au-S reaction as mass load. Thirdly, we dried the device at 37 °C for 30 min and hence 0.1 μg mass was loaded. The above adding-drying steps were repeated three times until a total of 0.3 μL of solution was added. The resonant frequencies were measured by an impedance analyzer (4294A, Agilent, Santa Clara, CA, USA) on the average resonant frequency values after three measurements. The relationship between resonant frequency and total mass load was thus obtained. The measured resonant frequencies were 79.965, 79.882, 79.803 and 79.681 kHz for 0, 0.1, 0.2, and 0.3 μg total loaded mass, respectively, as shown in Figure 5.

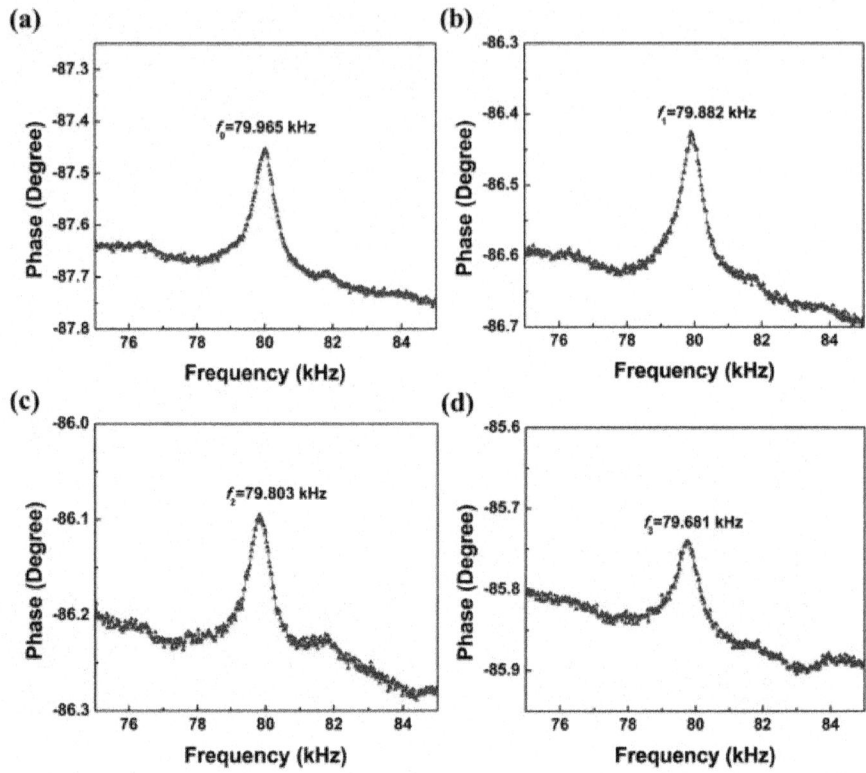

Figure 5: Resonant frequencies for different mass loaded: (a) no mass loaded; (b) 0.1 µg total mass loaded; (c) 0.2 µg total mass loaded and (d) 0.3 µg total mass loaded.

A nearly linear relationship between the frequency shift and total mass load of the device was found, as shown in Figure 6. From the results, the mass sensitivity can be calculated as S_m = 931 Hz/µg. Xu *et al.* [25] reported a biosensor array for immunoassay with a mass sensitivity of 6250 Hz/µg. Their piezoelectric diaphragm biosensors were based on PZT films and had smaller dimensions, which could increase the mass sensitivity. Zhao *et al.* [29] reported a lead-free piezoelectric biosensor based on polyvinylidene fluoride (PVDF) piezoelectric film with mass sensitivity of 185 Hz/µg, which was much smaller than our result.

The bio-sensing performance characteristics were also studied. In this test, the residual silicon thickness of the biosensor was about 20 µm. The detailed steps were as follows: (1) Thin gold film of 50 nm-thick was deposited on the backside of the diaphragm to serve as the immobilization layer, which was then cleaned and modified by piranha solution treatment; (2)

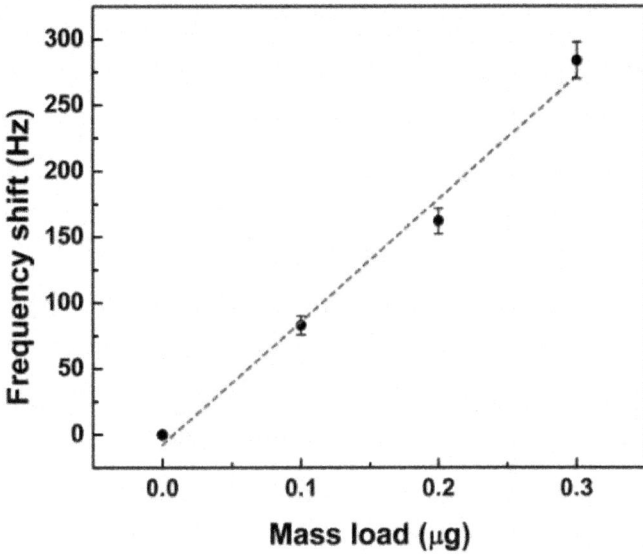

Figure 6: Relationship of the frequency shift and mass load.

The capture DNA probe with the same concentration and sequence, which was used in the test of mass sensitivity, was immobilized on the gold surface. After 6 h of immobilization, the reaction surface was washed by DI water and dried under N_2 flow; (3) 6-Mercapto-1-hexanol (MCH) blocking reagent was dropped for 2 h to improve the specificity. Then the reaction surface was washed by DI water and dried under N_2 flow; (4) In the last step, target analytes which contain complementary DNA with the concentration of 1 µg/µL was added on the reaction surface for 0.5 h of nucleic acid hybridization. Then the excess analyte were washed away by DI water and the device was dried under N_2 flow.

The resonant frequencies of the biosensor were measured immediately after each process, as shown in Figure 7. The original resonant frequency of the biosensor was noted as f_1, and the resonant frequencies after depositing gold film, adding probes, blockers and targets are f_2, f_3, f_4 and f_5, respectively. The resonant frequency for the plain, the deposition of gold, the immobilization of DNA probes, blockers and targets are 103.341, 102.635, 102.427, 102.176 and 101.988 kHz, respectively. The resonant frequency was continuously shifted to the lower domain, indicating an accumulated mass increasing during each process.

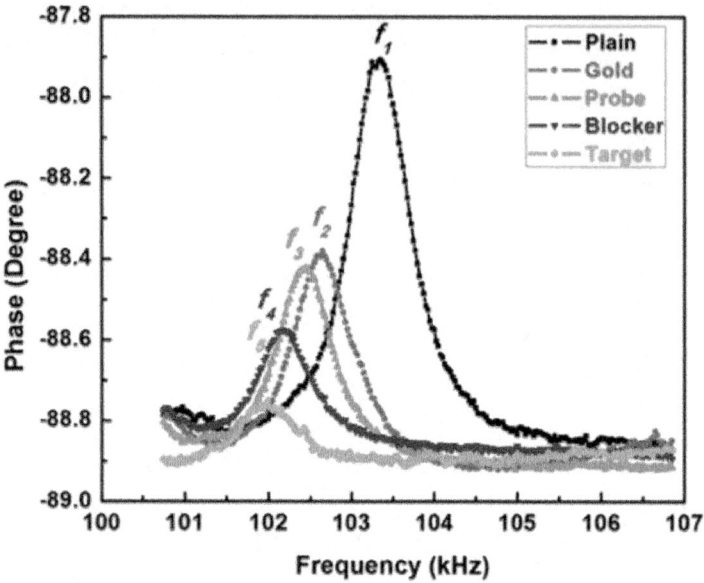

Figure 7: Frequency spectrum of the biosensor after different steps.

Detailed frequency changes after each process for the biosensor were shown in Figure 8, of which the data points were calculated based on the average values after three measurements. Comparing the frequency decreased values for the biosensor, $\Delta f_1 = f_1 - f_2$ is the largest, which may be due to the high density of gold film. The resonant frequency shift values for the immobilization of probes, blockers and targets are 0.208, 0.251 and 0.188 kHz, respectively.

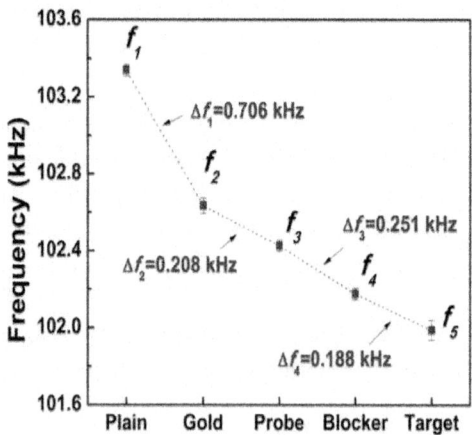

Figure 8: Frequency spectrum of the biosensor after different steps.

MATERIALS AND METHODS

Design of Device Structure and Sensing Principle

The schematic structure and principle of the device are plotted in Figure 9. The piezoelectric biosensor was mainly composed of a supporting layer, piezoelectric layer, electrode layer and immobilization layer. For this piezoelectric diaphragm biosensor, the piezoelectric layer is driven by applying the voltage through the converse piezoelectric effect. The resonant frequency of biosensor after fabrication is f_0, owing to a mass change of the diaphragm, the resonant frequency will decrease to f. The resonant frequency shift $(f - f_0)$ could be calculated, and thus the corresponding mass could be determined.

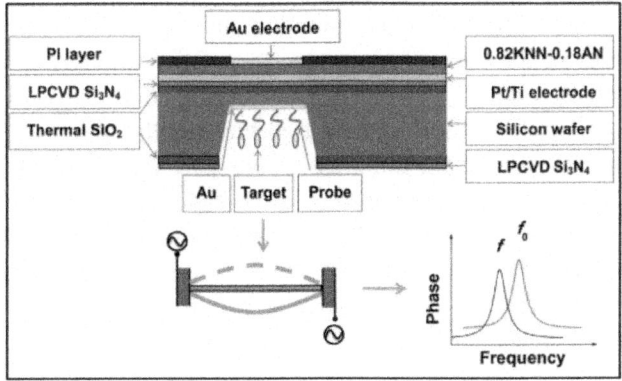

Figure 9: Schematic drawings and principle of the piezoelectric diaphragm biosensor.

The resonant frequency of a square diaphragm with length of a can be expressed as [25,28,31,32]:

$$f_0 = \frac{1}{2\pi a^2}\sqrt{\frac{\gamma^4 D}{\rho h}}$$

(1)

where f_0 is the resonant frequency, ρ is the diaphragm density, h is the diaphragm thickness, γ is a constant and D is the flexural rigidity, respectively. The definition of mass sensitivity of the biosensor is the resonant frequency change corresponding to a unit mass load, which can be defined as [25,33]:

$$S_m = -\frac{\Delta f}{\Delta m}$$

(2)

where Δm is the mass change per unit area, the Δf is the resonant frequency shift. A larger value of S_m means that the biosensor is more mass sensitive.

Fabrication of the Piezoelectric Diaphragm Biosensor

The piezoelectric diaphragms were fabricated by combining the chemical solution deposition method with traditional silicon micromachining technology; the fabrication processing steps are shown in Figure 10. In summary, the fabrication process was composed of nine main steps. (1) A layer of 1.5 μm-thick thermal SiO_2 was grown on a 4" double-sided polished silicon wafer; (2) A silicon nitride layer of 200 nm was then deposited by low pressure chemical vapour deposition (LPCVD) on both sides of the wafer; (3) The diaphragm windows on the backside were opened by dry etching of Si_3N_4 and wet etching of SiO_2, which were realized by inductive coupled plasma (ICP) using SF_6 + He and by buffered oxide etchant (BOE, which was composed of HF, NH_4F and H_2O), respectively; (4) The backside silicon was anisotropic wet etched by KOH until the remaining thickness of silicon was about 50 μm; (5) Pt/Ti layer of 200 nm/20 nm were sputtered on the front side of wafer as the bottom electrode; (6) 0.82KNN-0.18AN layer was deposited using the chemical solution deposition technique to serve as the piezoelectric layer; (7) A polyimide (PI) layer of 2 μm was spin-coated, patterned and cured as an insulation layer to minimize parasitic capacitance induced by the patterned electrode wiring; (8) Au layer of 80 nm to serve as top electrode was sputtered and patterned by lift-off process; (9) Finally, the backside silicon was etched off by ICP using SF_6 + O_2 + C_4F_8 until the required residual silicon thickness (20~25 μm) was reached.

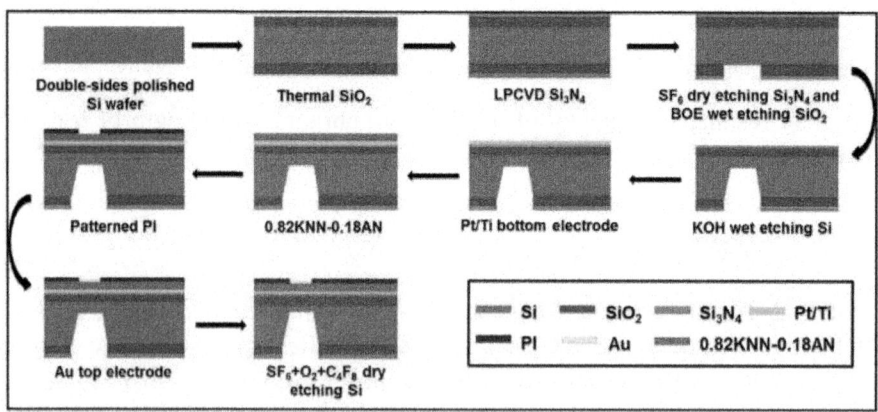

Figure 10: Schematic fabrication process of the piezoelectric diaphragms.

The thickness of the double-sided polished Si wafer was 400 μm and the crystal orientation was (100). Square diaphragms were fabricated by the anisotropic wet etching in 30 wt% KOH solution at 80 °C in the oil bath heater.

In the fabrication process, three lithography masks were used for the patterning of silicon diaphragms, the polyimide layer and the top electrodes, which were designed as shown in Figure 11. The alignment of the silicon diaphragm mask with the polyimide mask used the double-sided aligning technique while the polyimide mask with the top electrode mask used the overlay aligning technique.

(a)   **(b)**

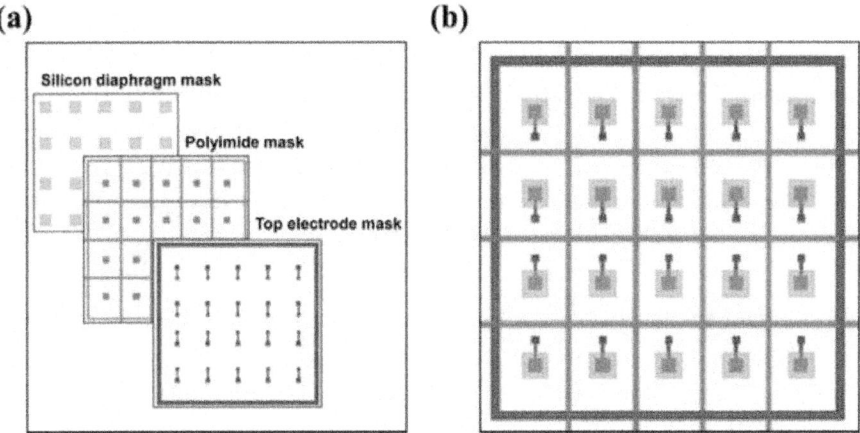

Figure 11: (a) Three masks for the fabrication process; and (b) their alignment image.

The lead-free 0.82KNN-0.18AN thin film layer was deposited using the chemical solution deposition method. For the preparation of the 0.82KNN-0.18AN precursor solution, potassium acetate, sodium acetate, silver acetate and niobium ethoxide were used as the starting chemicals and 2-methoxyethanol was used as the solvent. Polyvinylpyrrolidone (PVP) with a molecular weight of 360,000 was added to enhance the film thickness. The chemical composition was designed as $0.82K_{0.5}Na_{0.5}NbO_3$-$0.18AgNbO_3$. The final concentration of 0.82KNN-0.18AN precursor solution was 0.5 mol/L. The precursor solution was spin-coated onto the front side of the device at 2000 rpm for 60 s. Then wet film was dried at 100 °C for 2 min on a hotplate. The dried film was pyrolyzed at 330 °C for 5 min and annealed at 650 °C for 10 min in a rapid thermal annealing (RTA) furnace with a heating rate of 50 °C/s. Four layers were deposited by repeating all the above steps to increase the thickness of the 0.82KNN-0.18AN film.

Measurements of the Resonant Frequencies

The resonant frequencies were measured by the impedance analyzer (Agilent 4294A) at room temperature. The top electrode and bottom electrode were connected to the impedance analyzer using copper wires with a Agilent

42942A terminal adapter. During the measurements, the direct current (DC) bias voltage was fixed at 30 V and the oscillating (OSC) voltage level was 1 V.

CONCLUSIONS

In this paper, we have developed a novel lead-free piezoelectric diaphragm biosensor, unsing a fabrication process compatible with integrated circuit (IC) technology. The thickness of the piezoelectric film was about 2.16 µm. The resonant frequencies were measured and the frequency changes were calculated. The mass sensitivity of the biosensor was about 931 Hz/µg. After sensing performance characterization, it was verified that this piezoelectric biosensor can potentially be used for nucleic acid testing. With micro-machining technology and process optimization, the piezoelectric diaphragm biosensors can be miniaturized, so the cost can be further reduced.

ACKNOWLEDGMENTS

This work was financially supported by International Science & Technology Cooperation Program of China (Grant No. 2011DFA51880), the 111 Project of China (B14040), the Shaanxi Province International Collaboration Program (Grant No. 2012KW-02), the Shaanxi Science & Technology Promotion Program(2011TG-08), the Qianren Program of the Chinese Government, the Natural Science and Engineering Research Council of Canada (NSERC), and the United States Office of Naval Research (Grant No. N00014-12-1-1045). The SEM work was done at International Center for Dielectric Research (ICDR), Xi'an Jiaotong University, Xi'an, China; The authors also thank Yijun Zhang for his help in using SEM. The authors also thank Bei Zhao and Jie Hu for their help in DNA testing experiments.

AUTHOR CONTRIBUTIONS

All the authors conceived and designed the experiments. Xiaomeng Li performed the experiments; Xiaomeng Li and Xiaoqing Wu analyzed the data; Xiaoqing Wu, Peng Shi and Zuo-Guang Ye supervised and reviewed the manuscript.

REFERENCES

1.   Mehrabani, S.; Maker, A.J.; Armani, A.M. Hybrid Integrated Label-Free Chemical and Biological Sensors. *Sensors* 2014,*14*, 5890–5928.

2.   Kim, S.K.; Cho, H.; Park, H.J.; Kwon, D.; Lee, J.M.; Chung, B.H. Nanogap biosensors for electrical and label-free detection of biomolecular interactions. *Nanotechnology* 2009, *20*, 455502.

3. Esfandyarpour, R.; Javanmard, M.; Koochak, Z.; Esfandyarpour, H.; Harris, J.S.; Davis, R.W. Label-free electronic probing of nucleic acids and proteins at the nanoscale using the nanoneedle biosensor. *Biomicrofluidics* 2013, *7*, 044114.

4. Esfandyarpour, R.; Esfandyarpour, H.; Javanmard, M.; Harris, J.S.; Davis, R.W. Microneedle biosensor: A method for direct label-free real time protein detection. *Sens. Actuators B Chem.* 2013, *177*, 848–855.

5. Goda, T.; Miyahara, Y. Label-free and reagent-less protein biosensing using aptamer-modified extended-gate field-effect transistors. *Biosens. Bioelectron.* 2013, *45*, 89–94.

6. Goda, T.; Tabata, M.; Sanjoh, M.; Uchimura, M.; Iwasaki, Y.; Miyahara, Y. Thiolated 2-methacryloyloxyethyl phosphorylcholine for an antifouling biosensor platform. *Chem. Commun.* 2013, *49*, 8683–8685.

7. Miller, T.R.; Skoog, S.A.; Edwards, T.L.; Lopez, D.M.; Wheeler, D.R.; Arango, D.C.; Xiao, X.; Brozik, S.M.; Wang, J.; Polsky, R.; *et al.* Multiplexed microneedle-based biosensor array for characterization of metabolic acidosis. *Talanta*2012, *88*, 739–742.

8. Yaseen, M.T.; Yang, Y.C.; Shih, M.H.; Chang, Y.C. Optimization of High-Q Coupled Nanobeam Cavity for Label-Free Sensing. *Sensors* 2015, *15*, 25868–25881. [Google Scholar] [CrossRef]

9. Yang, Z.P.; Liu, X.; Zhang, C.J.; Liu, B.Z. A high-performance nonenzymatic piezoelectric sensor based on molecularly imprinted transparent TiO_2 film for detection of urea. *Biosens. Bioelectron.* 2015, *74*, 85–90.

10. Neves, M.A.; Blaszykowski, C.; Bokhari, S.; Thompson, S. Ultra-high frequency piezoelectric aptasensor for the label-free detection of cocaine. *Biosens. Bioelectron.* 2015, *72*, 383–392.

11. Kim, S.; Choi, S.J. A lipid-based method for the preparation of a piezoelectric DNA biosensor. *Anal. Biochem.* 2014, *458*, 1–3.

12. Tsai, J.Z.; Chen, C.J.; Shie, D.T.; Liu, J.T. Resonant efficiency improvement design of piezoelectric biosensor for bacteria gravimetric sensing. *Biomed. Mater. Eng.* 2014, *24*, 3597–3604. [Google Scholar] [PubMed]

13. Su, L.; Zou, L.; Fong, C.C.; Wong, W.L.; Wei, F.; Wong, K.Y.; Wu, R.S.; Yang, M. Detection of cancer biomarkers by piezoelectric biosensor using PZT ceramic resonator as the transducer. *Biosens. Bioelectron.* 2013, *46*, 155–161.

14. Alava, T.; Mathieu, F.; Rameil, P.; Morel, Y.; Soyer, C.; Remiens, D.; Nicu, L. Piezoelectric-actuated, piezoresistive-sensed circular

micromembranes for label-free biosensing applications. *Appl. Phys. Lett.* 2010, *97*, 093703.

15. Wang, Z.; Miao, J.; Zhu, W. Micromachined ultrasonic transducers and arrays based on piezoelectric thick film. *Appl. Phys. A* 2008, *91*, 107–117.

16. Wang, Z.; Miao, J.; Zhu, W. Piezoelectric thick films and their application in MEMS. *J. Eur. Ceram. Soc.* 2007, *27*, 3759–3764.

17. Laser, D.J.; Santiago, J.G. A review of micropumps. *J. Micromech. Microeng.* 2004, *14*, R35–R64.

18. Perçin, G.K.; Yakub, K.; Butrus, T. Micromachined droplet ejector arrays for controlled ink-jet printing and deposition.*Rev. Sci. Instrum.* 2002, *73*, 2193.

19. Hao, R.Z.; Song, H.B.; Zuo, G.M.; Yang, R.F.; Wei, H.P.; Wang, D.B.; Cui, Z.Q.; Zhang, Z.; Cheng, Z.X.; Zhang, X.E. DNA probe functionalized QCM biosensor based on gold nanoparticle amplification for Bacillus anthracis detection.*Biosens. Bioelectron.* 2011, *26*, 3398–3404.

20. Yao, C.; Zhu, T.; Tang, J.; Wu, R.; Chen, Q.; Chen, M.; Zhang, B.; Huang, J.; Fu, W. Hybridization assay of hepatitis B virus by QCM peptide nucleic acid biosensor. *Biosens. Bioelectron.* 2008, *23*, 879–885.

21. Sankaran, S.; Panigrahi, S.; Mallik, S. Olfactory receptor based piezoelectric biosensors for detection of alcohols related to food safety applications. *Sens. Actuators B Chem.* 2011, *155*, 8–18.

22. Johnson, B.N.; Mutharasan, R. Biosensing using dynamic-mode cantilever sensors: A review. *Biosens. Bioelectron.* 2012,*32*, 1–18.

23. Arora, P.; Sindhu, A.; Dilbaghi, N.; Chaudhury, A. Biosensors as innovative tools for the detection of food borne pathogens. *Biosens. Bioelectron.* 2011, *28*, 1–12.

24. Xiao, Y.; Liu, Y.; Borg, G.; Li, C.M. Design of a novel disposable piezoelectric co-polymer diaphragm based biosensor unit. *Mater. Sci. Eng. C* 2011, *31*, 95–98.

25. Xu, T.; Wang, Z.; Miao, J.; Yu, L.; Li, C.M. Micro-machined piezoelectric membrane-based immunosensor array.*Biosens. Bioelectron.* 2008, *24*, 638–643.

26. Hwang, I.H.; Lee, J.H. Self-actuating biosensor using a piezoelectric cantilever and its optimization. *J. Phys. Conf. Ser.*2006, *34*, 362–367.

27. Nicu, L.; Guirardel, M.; Chambosse, F.; Rougerie, P.; Hinh, S.; Trevisiol, E.; Francois, J.; Majoral, J.; Caminade, A.; Cattan, E.; *et al.* Resonating piezoelectric membranes for microelectromechanically based bioassay: Detection of streptavidin-gold nanoparticles interaction with biotinylated

DNA. *Sens. Actuators B Chem.* 2005, *110*, 125–136.

28. Olfatnia, M.; Singh, V.R.; Xu, T.; Miao, J.M.; Ong, L.S. Analysis of the vibration modes of piezoelectric circular microdiaphragms. *J. Micromech. Microeng.* 2010, *20*, 085013.

29. Zhao, B.; Hu, J.; Ren, W.; Xu, F.; Wu, X.; Shi, P.; Ye, Z. A new biosensor based on PVDF film for detection of nucleic acids. *Ceram. Int.* 2015, *41*, S602–S606.

30. Li, X.; Wu, X.; Ren, W.; Shi, P.; Ye, Z. Preparation and characterization of sodium potassium niobate-silver niobate lead-free films by chemical solution deposition. *Ceram. Int.* 2015, *41*, S228–S233.

31. Xu, T.; Miao, J.; Wang, Z.; Yu, L.; Li, C.M. Micro-piezoelectric immunoassay chip for simultaneous detection of Hepatitis B virus and α-fetoprotein. *Sens. Actuators B Chem.* 2011, *151*, 370–376.

32. Olfatnia, M.; Xu, T.; Ong, L.S.; Miao, J.M.; Wang, Z.H. Investigation of residual stress and its effects on the vibrational characteristics of piezoelectric-based multilayered microdiaphragms. *J. Micromech. Microeng.* 2010, *20*, 015007.

33. Xin, Y.; Li, Z.; Odum, L.; Cheng, Z.Y.; Xu, Z. Piezoelectric diaphragm as a high performance biosensor platform. *Appl. Phys. Lett.* 2006, *89*, 223508.

Chapter 3

DESIGN AND ANALYSIS OF A SENSOR SYSTEM FOR CUTTING FORCE MEASUREMENT IN MACHINING PROCESSES

Qiaokang Liang[1,2], Dan Zhang[3], Gianmarc Coppola[3], Jianxu Mao[1], Wei Sun[1,2], Yaonan Wang[1,2] and Yunjian Ge[4]

[1]College of Electric and Information Technology, Hunan University, Changsha 410082, Hunan, China

[2]State Key Laboratory of Advanced Design and Manufacturing for Vehicle Body, Hunan University, Changsha 410082, Hunan, China

[3]Lassonde School of Engineering, York University, Toronto, ON M3J 1P3, Canada

[4]Institute of Intelligent Machines, Chinese Academy of Science, Hefei 230031, Anhui, China

ABSTRACT

Multi-component force sensors have infiltrated a wide variety of automation products since the 1970s. However, one seldom finds full-component sensor systems available in the market for cutting force measurement in machine processes. In this paper, a new six-component sensor system with a compact monolithic elastic element (EE) is designed and developed to detect the tangential cutting forces Fx, Fy and Fz (*i.e.*, forces along x-, y-, and z-axis) as well as the cutting moments Mx, My and Mz (*i.e.*, moments about x-, y-, and z-axis) simultaneously. Optimal structural parameters of the EE are carefully designed via simulation-driven optimization. Moreover, a prototype sensor system is fabricated, which is applied to a 5-axis parallel kinematic machining center. Calibration experimental results demonstrate that the system is capable of measuring cutting forces and moments with good linearity while minimizing coupling error. Both the Finite Element Analysis (FEA) and calibration experimental studies validate the high performance of the proposed sensor system that is expected to be adopted into machining processes.

INTRODUCTION

Computer Numerical Control (CNC) machines are operated by precisely programmed commands and widely used in modern manufacturing industry, enabling the manufacture of complex-shaped products or parts that cannot be produced by manually operated machines.

Nowadays, every manufacturer in the global market is pitted against worldwide competitors with consistently improving product quality, enhanced manufacturing productivity, elimination of inspections, and shrinking total machining costs.

The condition of cutting tools and the cutting process should be identified without human assistance and/or interrupting the manufacturing process operation, which is one of the most important operating criteria that influences the manufacturing quality and productivity. An automated machining process with a successful machining condition monitoring system could allow increased principal equipment utilization, hence achieving considerable cost savings for the manufacturing industry [1,2].

As precise estimation and determination of cutting force at tool tips in manufacturing processes is the most effective method for monitoring machine tools and the machining processes, it is indispensable in many research and application fields for monitoring the tool conditions, analyzing of machining methods and tools, estimating real-time tool wear [3], designing proper machining tools, characterizing and optimizing manufacturing process and cutting parameters, *etc.*Consequently, sensor systems for measuring cutting force are receiving widespread attention within the equipment manufacturing and other related industries community.

The cutting forces during machining process are dependent on many parameters such as depth of cut, configuration of the cutting tool, feed rate, material of the workpiece and the tool, and some unknown factors such as cooling method. Therefore, theoretical cutting force calculations are hardly applicable to monitor practical machining processes in a factory environment.

A great number of sensor systems with many sophisticated signal and information processing techniques proposed for the pertinent applications can be found in the literature. Additionally, a lot of commercial devices capable of measuring the cutting force during milling, turning, drilling or grinding have been developed by many companies such as Kistler TLC (Technische Lehrmittel Construktion, Unna, Germany), *etc.*

Conveniently, the cutting force measurement systems could be separated into at least two groups, consisting of indirect and direct approaches. The direct measurement approach refers to detecting and quantifying the cutting force by using force dynamometers directly. This is done by relying on some specific effects such as mechano-magnetic, mechano-electric, and mechano-optic conversions, while the indirect measuring approach refers to a cutting force that is estimated by dynamic parameters associated with cutting force such as acceleration, acoustic emission, vibration, and spindle motor current. The most common indirect approaches to cutting force measurement or monitoring are based upon motor current and power measurement [4,5]. Shin *et al.* [6] designed an indirect approach for measuring cutting force in end-milling process with a three-axis acceleration sensor and Hall-effect current sensors. An indirect approach for detecting cutting force for an Adaptive Control with Constraints (ACC) system for machining centers by using the currents of a.c. feed-drive servo drives was reported in [7]. A novel approach for measuring cutting force via command voltages drawn by electro-magnetic bearing was proposed by Auchet *et al.* [8], and the experimental results show that the calculated cutting forces are in good agreement with commercial Kistler measurement system. The static accuracy and bandwidth of the Altintas *et al.* presented a comprehensive model that evaluates the cutting force in normal and tangential directions through modeling the cutting mechanics, feed variation and tool indentation effect [9]. In recent years, different methods for evaluating or predicting cutting forces based on tool indentation effect [10], nonlinear cutting force coefficients [11], and rotational DOF (Degrees of Freedom) of the multi-axis kinematics effect on mechanics of the cutting process [12] are considerably attracting the attentions of researchers. Within most environments the indirect approach is easier to achieve, but has several limitations such as high computation time, unsuitable for multi-axis cutting process, does not consider the frictional behavior of the machine tools.

On the other hand, direct measuring approaches have been widely employed because they can provide more accurate measurement of cutting forces. Although, they have their limitations like high cost, layout constraints, fragility to overload,*etc.* Rao, Gao, and Friedrich presented an integrated force measurement method for on-line cutting geometry inspection, and a piezoelectric force sensor with high resolution (0.44 mN) and sensitivity (7 mV/gm) which was adopted to measure the feed force combined with thrusts during the diamond-cutting process [13]. A novel cylindrical capacitive displacement sensor and a magnetic exciter for monitoring end milling processes were proposed by Kim *et al.* [14]. It was verified by cutting experiments that the proposed sensory system was suitable in the monitoring of high speed cutting conditions on various machine tools. A strain gauge based tool dynamometer and a piezo-film accelerometer, with a natural frequency of 2.35 kHz, was developed to detect dynamic and static cutting force for ultra-precision machine tools [15]. Jin, Venuvinod, and Wang proposed an approach for detecting cutting force with an optical fiber sensor, and the results of calibration experiment and manufacturing tests show the proposed system obtains satisfactory performances (Its static sensitivity and linearity of the system are 2.51 mV/N and 1.2% Full Scale (F.S.), respectively, with a natural frequency of 950 Hz) [16]. The pros and cons as well as typical designs of the mentioned intrinsic transduction techniques are summarized in the Table 1. These transduction techniques are commonly interconnected in practical cutting forces measurement systems [17,18].

Table 1: Intrinsic transduction techniques for cutting force measurement

Measuring Technology	Direct/Indirect Measurement	Pros	Cons	Typical Designs
Current	indirect	• easier to achieve • cost effective	• time-consuming • unsuitable for multi-axis cutting process • without consideration of the frictional behavior of the machine tools	[5]
Voltage	indirect	• wide bandwidth (up to 4 kHz) • easy to conversion and processing	• Limited to stable conditions • susceptible to electromagnetic interference	[8]
Strain gauge	direct	• simple construction • high and adjustable resolution • high reliability	• higher power consumption • rigid and fragile • scarce reproducibility	[15]
Capacitive	direct	• high sensitivity and resolution • long-time stability • Adaptability to Environment	• temperature sensitive • stray capacitance • Edge effect	[14]
Optoelectronic	direct	• good reliability • wide measurement range • good adaptability to workshop conditions	• non-conformable • hard to construct dense arrays	[16]
Piezoelectric	direct	• high frequency response and high dynamic range • rangeability • higher accuracy and finer resolution • high sensitivity and stiffness	• charge leakages • poor spatial resolution • deteriorations of voltages or drifts in the presence of static forces	[13]

Many different sensory systems for cutting force measurement have been developed for specific tasks. However, many tasks such as monitoring and controlling of high speed machining processes require reliable and practical sensing methods to measure the cutting force, which cannot be achieved by the majority of the present sensory systems. This is partially due to limited workpiece sizes, low frequency bandwidths, mounting constraints, wiring complexities, and susceptibility to harsh machining environments [19]. In this paper, a novel six-component Force/Moment (F/M) sensor is designed to detect the normal and tangential cutting force (Fx, Fy and Fz), as well as the cutting moments (Mx, My and Mz), simultaneously.

SYSTEM CONFIGURATION

The experimental setup for implementing the sensor system for cutting force measurement is shown in Figure 1. The six-component cutting F/M sensor is sandwiched between the adapter and the moving platform of a 5-axis parallel kinematic machining center. After calibration, the multi-component F/M sensor is capable of measuring the normal and tangential cutting force (Fx, Fy and Fz), as well as the cutting moments (Mx, My and Mz), simultaneously.

Raw signals from the sensor are processed, stored and transmitted to the host-computer by the integrated electrical circuit board as shown in Figure 2. Feature extraction of signals with the appropriate signal processing algorithms is adopted to obtain a simplified signal and specific characteristic features related to the cutter conditions. Decision-making strategies analyze the characteristic features and perform a pattern association task. Artificial neural networks, fuzzy logic system, expert systems, and other hybrid intelligent systems are proving their effectiveness in decision making of condition monitoring systems to make sure efficient removal of metals and taking correct action to prevent harm in the event of breakages or accidents. The input signals were also used to modify the part of the program that is used in the second pass with alternative tool path, feed rate, cutting speed, stock removal.

Figure 1: Experimental setup for implementing the sensor system for cutting force measurement.

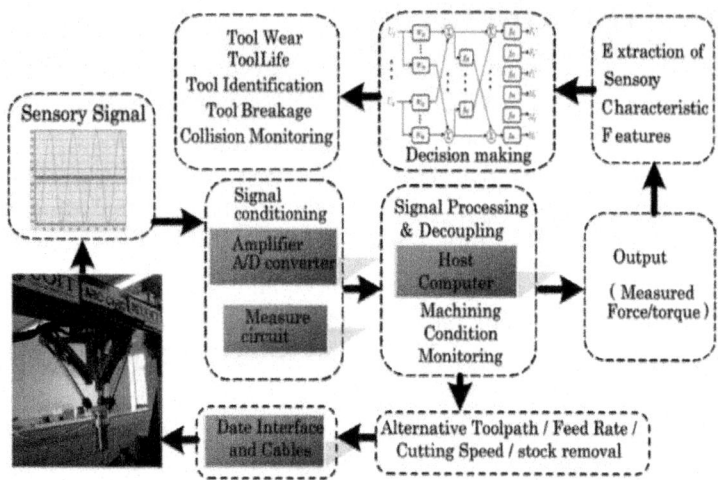

Figure 2: Schematic illustration of strategy of the system.

DESIGN AND CONSTRUCTION OF THE SENSOR

The most common technique to detect cutting F/M relies on resistive measurement approach with strain gauges. This is due to its advantages such as high reliability with time, excellent linearity over wide range, simplicity, wide operating temperature range, adjustable resolution, maintenance free and relatively low in cost, *etc.* [20]. The sensing system based on resistive approach consists of four functional elements: strain gauges, Elastic element (EE) with a specially designed shape, measuring bridges, and electric circuit.

Sensing Principle

The strain gauges bonded onto the EE are sensitive to dimensional changes when a cutting force is applied to the sensor, and their resistances change according to:

$$\frac{\Delta R_i}{R_i} = G\varepsilon_i \tag{1}$$

where $\Delta R_i/R_i$ is the variation of relative resistance ratio of the ith strain gauge. G and ε_i are the gauge factor of the strain gauge and specific strain, respectively. Provided that the EE behaves within the elastic range of the material, the strains and the applied cutting load are related by the following equation:

$$\varepsilon_i = f_i(\mathbf{F}) \tag{2}$$

where $F \in R^6$ is the cutting load vector that contains the forces $\mathbf{F}_s$ and the moments $\mathbf{M}_s$ applied onto the sensor.

Wheatstone Bridge arrangements are generally employed as measuring bridges via electrically connecting several strain gauges (generally 4, 8, or a still greater number) to measure the small resistive changes with high sensitivity and inherent linearity. Therefore, there is a relation between the output voltage variation of the measuring bridges and the strains due to the applied cutting load:

$$\mathbf{\Delta V_0} = V_E \mathbf{GW\varepsilon} = \mathbf{TF}$$

(3)

where $\Delta V_0 \in R^6$, V_E is the voltage excitation source of the bridges, $W \in R^{6 \times 24}$ and $T \in R^{6 \times 6}$ represent the bridge transformation matrix and the transducer matrix, respectively. The transducer matrix is a constant matrix that depends on the configuration of the measuring bridges, the structure of the EE, and the specific location of the strain gauges on the EE. After calibration or structural analysis, the applied cutting load can be determined and calculated:

$$\mathbf{F} = \mathbf{T}^\# \mathbf{\Delta V_0} + (\mathbf{I} - \mathbf{T}^\# \mathbf{T})\mathbf{z}$$

(4)

where $T^\#$ is the generalized inversion of $\mathbf{T}$. $\mathbf{z}$ is an arbitrary 6×1 vector, and usually set as $\mathbf{z} = \mathbf{o}$.

Multi-component cutting F/M sensor should be designed to be equally sensitive or accurate among its all force and moment components. As a consequence, the sensor and its electrical circuit should be characterized with simple but effective decoupling methods, identical amplification among components associated with a considerable degree of integration. The transducer matrix is normalized to avoid the unit inconsistency problem:

$$\mathbf{\overline{T}} = \mathbf{N}_{VN}^{-1} \mathbf{T} \mathbf{N}_{FN}$$

(5)

with normalization compensates:

$$\mathbf{N}_{VN} = \mathrm{diag}\left\{\Delta V_{1M}, \Delta V_{2M}, \Delta V_{3M}, \Delta V_{4M}, \Delta V_{5M}, \Delta V_{6M}\right\}$$

(6)

$$\mathbf{N}_{FN} = \mathrm{diag}\left\{F_{xM}, F_{yM}, F_{zM}, M_{xM}, M_{yM}, M_{zM}\right\}$$

(7)

where ΔV_{iM} and F_{iM} (or M_{iM}) are maximal voltage variation of the ith measuring bridge and pre-specified maximal cutting forces (or moments). The anisotropy index of the sensor can be obtained by the condition number of $\mathbf{\overline{T}}$ [21]:

$$C_o(\overline{\mathbf{T}}) = \frac{\sigma_{\max}(\overline{\mathbf{T}})}{\sigma_{\min}(\overline{\mathbf{T}})}$$

(8)

where $\sigma_{\max}$ and $\sigma_{\min}$ represent the largest and the smallest singular values of $\overline{\mathbf{T}}$, respectively.

The absolute sensitivity of the sensor can be evaluated by [22]:

$$S_o = \text{trace}(\overline{\mathbf{T}}^T\overline{\mathbf{T}}) = \sum_{i=1}^{6}\sigma_i^2$$

(9)

or:

$$S_o = \det(\overline{\mathbf{T}}^T\overline{\mathbf{T}}) = \prod_{i=1}^{6}\sigma_i$$

(10)

The obtained cutting load is respect to the sensor's coordinate frame $\{\mathbf{S}\}$, and should be transformed to the tool's coordinate frame $\{\mathbf{T}\}$ as:

$$\mathbf{F_t} = \mathbf{J_t}\mathbf{F}$$

(11)

where:

$$\mathbf{J_t} = \begin{bmatrix} \mathbf{R_t} & \mathbf{O} \\ \mathbf{S(r_t)R_t} & \mathbf{R_t} \end{bmatrix}$$

(12)

Here, $\mathbf{r}_t$ is the location vector of the frame $\{\mathbf{S}\}$ with respect to the frame $\{\mathbf{T}\}$, $\mathbf{R}_t$ is the is the orientation matrix of frame $\{\mathbf{S}\}$ relative to the frame $\{\mathbf{T}\}$, $\mathbf{O}$ is the null matrix, and $\mathbf{S(r}_t)$ represents skew-symmetric matrix operator of vector $\mathbf{r}_t$.

Structure of the Proposed Six-Component Cutting F/M Sensor

The distinctive design of the proposed six-component cutting F/M sensor consists of four parts as illustrated in Figure 3. In particular, the upper and lower adapters are connected to the upper and lower portions of the EE by stainless steel screws with controlled torque, respectively. The integrated electric circuit is mounted in the EE, and electrically connected to host computer via the electric connector. The sensor is mounted to the spindle and the machine center via the upper and lower adapters, respectively.

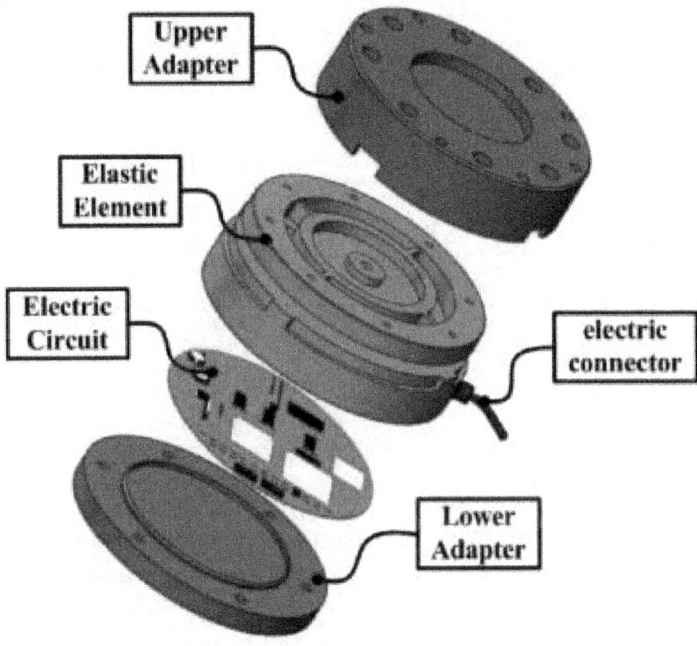

Figure 3: Structure of the proposed sensor.

Structure of Elastic Force-Sensing Elements

Briefly, the monolithic EE consists of double circular diaphragms, connected by a center hollow cylinder, four lamellas whose axes are perpendicular to each other in a cross-shape, a load-transmitting circular loop that transmits the cutting load, and a base frame (as indicated in Figure 4). The base frame and the circular loop connect with the upper and lower adapters by bolts, respectively.

Practically, the EE transmits the applied cutting load into deformations of its flexible sensing portions through its double diaphragms and lamellas. In particular, the upper diaphragm that serves as an elastic portion is bonded with two groups of strain gauges, and it is responsive to the applied moment to be measured Mx, My. Similarly, the lower diaphragm that serves as an elastic portion is bonded with three groups of strain gauges, and it is responsive to the applied force Fx, Fy, and Fz. Additionally, the lamellas that act as elastic portions are bonded with four strain gauges, and they are sensitive to the moment Mz.

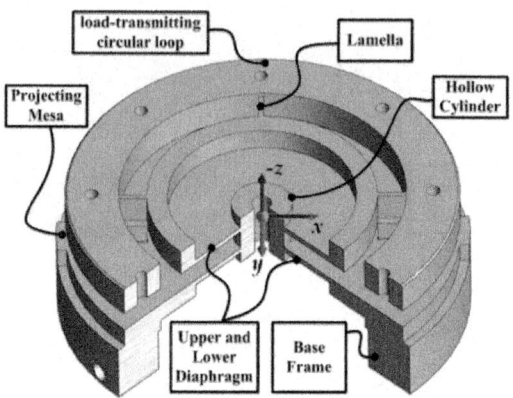

Figure 4: 3D model of the monolithic EE structure.

When the forces to be measured Fx, Fy or Fz act on the sensor, the circular loop connected with the upper adapter floats, while the base frame conneted with the lower adapter acts as a fixed support. Consequently, corresponding elastic deformation and strain will take place on the lower diaphragm. Accordingly, when the moment to be measured Mx (or My) act on the sensor, elastic deformations mainly take place on the upper and lower circular diaphragms. When the moment to be measured, Mz, acts on the load-transmitting circular loop, the inner borders of the lamellas provide fixed support while the outer borders of the lamellas float. Corresponding elastic deformation and strain consequently take place on the lamellas.

OPTIMAL DETERMINATION OF THE ELASTIC ELEMENT DIMENSIONS

Theoretical Analysis

The radial and tangential strains on the diaphragms induced by the applied cutting force can be expressed by:

$$\xi_r(r) = -\frac{h}{2}\frac{d^2\zeta}{dr^2} = \frac{-3F(\varsigma+2)(1-\mu^2)}{4\pi Eh^2} \tag{13}$$

$$\xi_t(r) = -\frac{h}{2}\frac{d\zeta}{dr} = \frac{-3\varsigma F(1-\mu^2)}{4\pi Eh^2} \tag{14}$$

where:

$$\varsigma = 2\ln\frac{2r}{d} - \iota - \frac{\iota d^2}{4r^2}$$
(15)

$$\iota = \frac{4D^2}{d^2 - D^2}\ln\frac{D}{2d}$$
(16)

where E and μ are the Young's modulus and the Poisson's ratio, respectively. d, D, and h are the inner circle diameter, the extended circle diameter, and the thickness of the diaphragm, respectively.

The corresponding radial and tangential stresses that occur on the diaphragm can be obtained by the expressions below:

$$\vartheta_r(r) = \frac{E}{1 - \mu^2}(\xi_r + \mu\xi_t) = \frac{-3F}{4\pi h^2}[(\mu + 1)\varsigma + 2]$$
(17)

$$\vartheta_t(r) = \frac{E}{1 - \mu^2}(\xi_t + \mu\xi_r) = \frac{-3F}{4\pi h^2}[(\mu + 1)\varsigma + 2\mu]$$
(18)

The radial strain and stress on the lamellas (of dimensions l, b, and h) induced by the applied cutting force can be expressed by:

$$\xi_r(x) = -\frac{h}{2}\frac{d^2\zeta}{dx^2} = \frac{6F}{bh^2E}\left(\frac{3}{2}l - x\right)$$
(19)

$$\xi_t(x) = -\frac{Eh}{2}\frac{d^2\zeta}{dx^2} = \frac{6F}{bh^2}\left(\frac{3}{2}l - x\right)$$
(20)

Simulation-Driven Optimization (SDO)

In general, multi-component force sensors always suffer from a critical drawback in the form of tradeoffs. These tradeoffs are typical in mechanical design and consist of a compromise between different performance characteristics such as sensitivity and stiffness, isotropy and coupling, static and dynamic performances. For example, optimum absolute sensor sensitivity corresponds to the desire of generating as much elastic strain as possible in the EE. However, both sensor sensitivity and general stiffness strongly depend on the geometrical parameters of the EE structure. More specifically, maximizing the sensor sensitivity is in contradiction with maximizing the general stiffness. Consequently, the optical design and development of a multi-component cutting F/M sensor is complicated and cannot be realized through direct optimization methods.

The SDO makes the procedure of new product development faster and more efficient. Design Exploration (DE) is a useful tool supplied by ANSYS Inc. (Pittsburgh, PA, USA), which could be performed to simulate the part response according to input parameter changes. Moreover, associated with response surface tools and optimization approaches, it can provide the most appropriate choices of input parameters to fulfill the system requirements. A design parameters study was performed via SDO to yield a set of feasible complete designs and make sure that the proposed sensor meets given performance requirements. A set of design parameters is defined by the independent geometric dimensions of the lamellas, the upper and lower diaphragms. Additionally, the max-deformation, the min-strain, the max-stress, and the primary response are set as output parameters. Table 2 shows the ranges for design parameters and objectives for output variables considered in the design parameter study. Each objective is set with the identical importance weight. Specifically, the first and third objectives about deformation and stress are adopted to enable that the EE works in the region below its elastic limit and at the same time enables cutting force measurement with satisfying stiffness, accuracy and linearity. Additionally, the second and fourth objectives related to the strain and the primary response frequency are considered to make sure that the system has superior performances such as high resolution and sensitivity, as well as satisfying dynamic performance.

Table 2: Design parameters and output variables

	Design Parameters				Output Variables			
Parameter	Extended Circle Diameter	Diaphragm Thickness	Lamella Thickness	Lamella Length	Maximum Total Deformation (mm)	Minimum Elastic Strain (10^{-6} mm/mm)	Maximum Equivalent Stress (MPa)	The Primary Response Frequency (Hz)
Variable	P1	P2	P3	P4	P5	P6	P7	P8
Bound/Objective	$40 < P1 < 50$	$1 < P2 < 4$	$1 < P3 < 3$	$7 < P4 < 12$	$P5 < 0.03$	$P6 > 100$	$P7 < 60$	$P8 > 500$
Candidate 1	49.55	1.055	11.119	1.378	0.0273	159.884	39.572	744.3
Candidate 2	46.35	1.031	7.334	2.402	0.0265	134.272	54.719	1210.4
Candidate 3	47.95	1.149	10.297	2.658	0.0192	130.072	40.916	1043.7

Figure 5 shows a graphical display of design parameters that demonstrate particular configurations of EE model associated with the corresponding output variables. Additionally, it also illustrates what each objective could achieve and if it entails sacrificing the objective of another output variable. Three best candidate sets with corresponding output variables that are most in alignment with all objectives are listed in the Table 2. Candidate set 1 is adopted in this design.

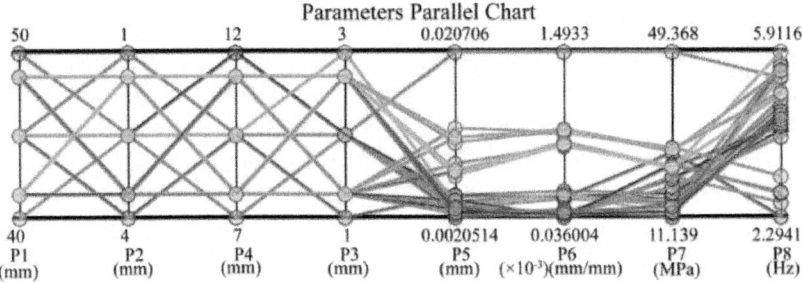

Figure 5: Parameters chart of SDO: design points with different colors generated via varied combination of the input parameters.

The influence ranks of input parameters over the output variables are calculated and shown in the Figure 6, and the greater positive sensitivity indicates greater impact on output variables positively, while more negative sensitivity means greater negative impact on output variables. From the sensitivity plot, it can be seen that the most noticeable parameter to each output variable are the membrane thickness as well as the lamella thickness.

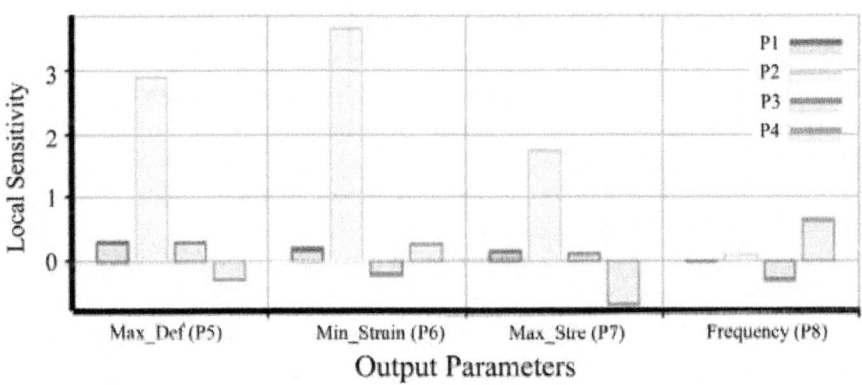

Figure 6: Sensitivity of the input parameters with respect to each output variables.

The relationships between design parameters and the variation of the output variables are illustrated in Figure 7. One can view the sensing system performances vary with respect to the design parameters intuitively, which is helpful to identify and understand specific changes to meet the corresponding requirements for the sensing system.

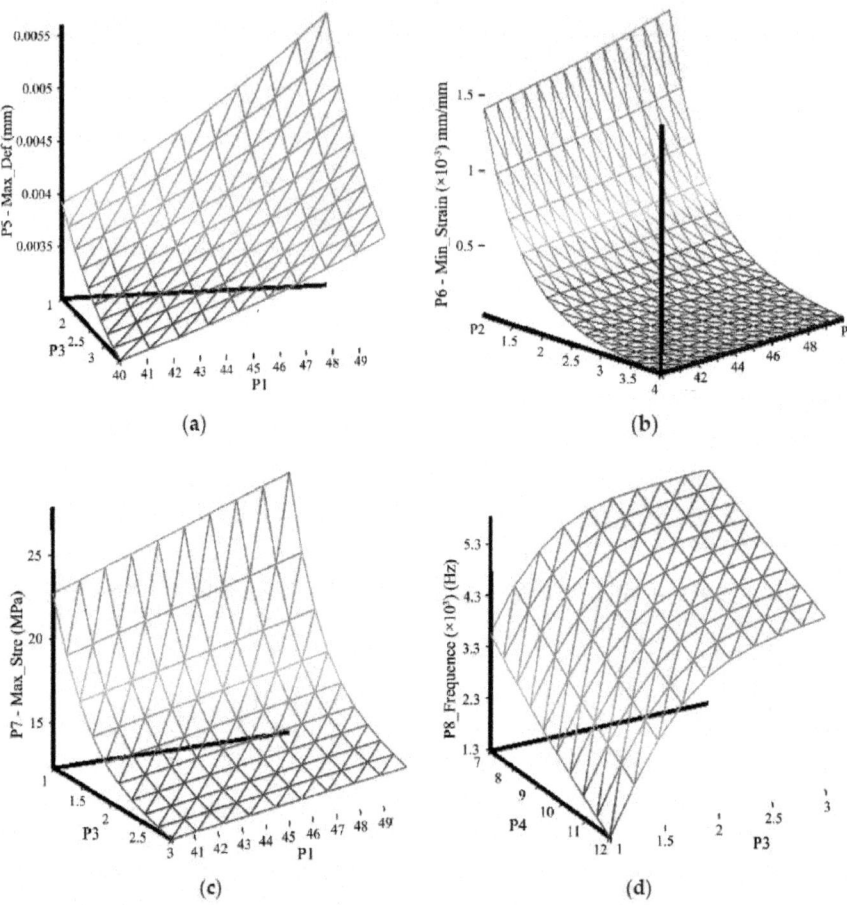

Figure 7: Relationships between design parameters and output variables: (a) Maximum deformation *versus* P1 and P3; (b) Minimum strain *versus* P1 and P2; (c) Maximum equivalent stress *versus* P1 and P3; (d) The first response frequency *versus* P4 and P5.

Dynamic Properties of the Sensor

The cutting force sensing system will respond dynamically when it is subject to applied cutting force in machining processes. Therefore, the identification of dynamic properties of the sensor is important both for dynamic measurement and control implementation thereafter.

A modal analysis is performed to measure and analyze the dynamic response of the sensor, e.g., vibration characteristics. A stiff support is set to the sensor base frame, and the interest responding frequencies is limited from the first to sixth frequencies, which are evaluated and displayed in Table 3. Moreover, the result also illustrate that the first two natural frequencies are almost the identical due to the symmetric structure of the sensor.

Table 3: The first six natural frequencies of the sensor

Mode	1	2	3	4	5	6
Responding Frequency (Hz)	744.3	748.4	1291.8	1317.7	2869.9	2999.5

Furthermore, a harmonic analysis is performed with ANSYS to identify and predict the sustained dynamic behavior of the sensor subjected to harmonically varying loads. When a load and a fixed support are applied at the circular loop and base frame of the sensor(see Figure 4) respectively, the corresponding output normal elastic strain of the diaphragms are obtained (as illustrated in Figure 8) to verify whether the design can successfully withstand forced vibration and not undergo resonance.

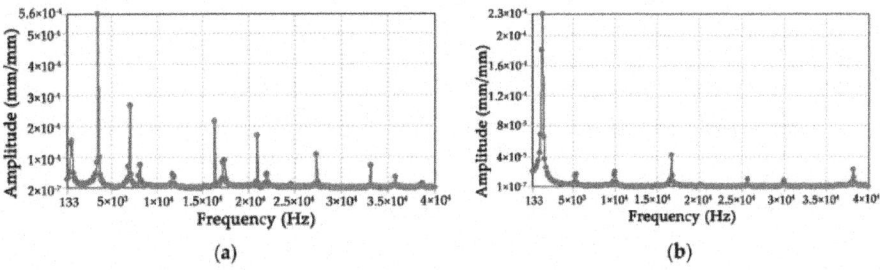

(a)   (b)

Figure 8: Harmonic response of the diaphragm under the measuring force component Fx (a) and Fz (b).

CUTTING FORCE SENSOR FABRICATION

The Arrangement Strategy of the Strain Gauges

Six groups of strain gauges are mounted in the areas of highest strain on the EE with considerations of maximizing the sensitivity to various components of cutting force and moment. 24 strain gauges (LY11, manufactured by HNM GmbH Inc., Balgheim, Germany) are bonded on the EE as shown in the Figure 9. Specifically, active strain gauges R1, R2, R3, and R4 (R5, R6, R7, and R8) are arranged on the lower diaphragm along the x-axis (y-axis) to detect force Fx (force Fy). Another four 45-degree active gauges R9, R10, R11, and R12 arc bonded on the lower diaphragm along its radial axis to measure cutting force along z-axis Fz. Also, four active strain gauges R13, R14, R15, and R16

(R17, R18, R19, and R20) are instrumented on the upper diaphragm along the x-axis (y-axis) to detect cutting moment about y-axis My (cutting moment about x-axis Mx). Additionally, four active gauges R21, R22, R23, and R24 are bonded on the lamella along its longitudinal direction to detect cutting moment about z-axis Mz.

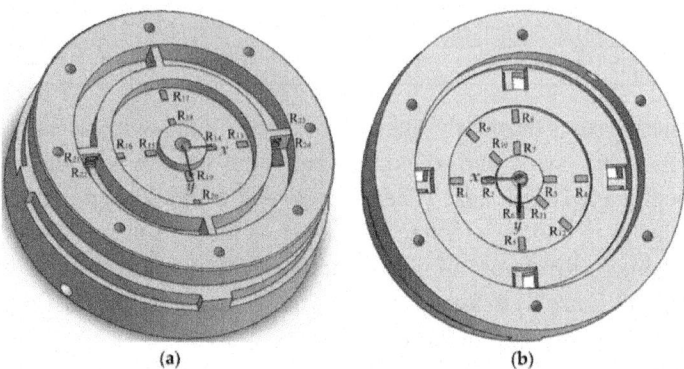

(a)　　　　　　　　　　　　　　　(b)

Figure 9: Arrangement strategy of the strain gauges: (a) Gauges emplacement on the upper diaphragm; (b) Gauges emplacement on the lower diaphragm.

The bonded gauges resistances will vary when the external cutting forces are applied to the sensor. Table 4 summarizes the variations of the each gauge resistance due to six components of the cutting force with the FEA. In addition, the symbols "+" and "−" in the table indicate an increase and decrease of the gauge resistance, respectively. While the symbols "=" and "0" represent identical and no variation, respectively. It is revealed that the strains obtained from the derived equations are in agreement with the results from the FEA.

Table 4: Variations of resistances of strain gauges

		Fx	Fy	Fz	Mx	My	Mz			Fx	Fy	Fz	Mx	My	Mz
Fx-bridge	R_1	+	0	+	0	0	+	Mx-bridge	R_{17}	0	+	+	+	0	+
	R_2	−	0	−	0	0	+		R_{18}	0	+	−	−	0	+
	R_3	+	0	−	0	0	+		R_{19}	0	−	−	+	0	+
	R_4	−	0	+	0	0	+		R_{20}	0	−	+	−	0	+
Fy-bridge	R_5	0	+	+	0	0	+	My-bridge	R_{13}	−	0	+	0	+	+
	R_6	0	−	−	0	0	+		R_{14}	−	0	−	0	−	+
	R_7	0	+	−	0	0	+		R_{15}	+	0	−	0	+	+
	R_8	0	−	+	0	0	+		R_{16}	+	0	+	0	−	+
Fz-bridge	R_9	+	+	+	0	0	+	Mz-bridge	R_{21}	=	=	+	0	+	−
	R_{10}	−	−	−	0	0	+		R_{22}	=	=	+	0	+	+
	R_{11}	+	+	−	0	0	+		R_{23}	=	=	+	0	+	+
	R_{12}	−	−	+	0	0	+		R_{24}	=	=	+	0	+	+

The Measuring Circuit

In practice, strain gauges measurements always involve quantities smaller than a few milli-strain (10^{-3} mm/mm), and their corresponding changes in resistance are extremely minute. Therefore, strain gauges are electrically connected to form six full-bridge configuration Wheatstone bridge circuits to detect six components of cutting load with high accuracy independently, as shown in Figure 10. The excitation diagonals of the Wheatstone bridges are connected to a voltage excitation, and the corresponding outputs voltage ΔVi will appear on the measurement diagonals of the bridges.

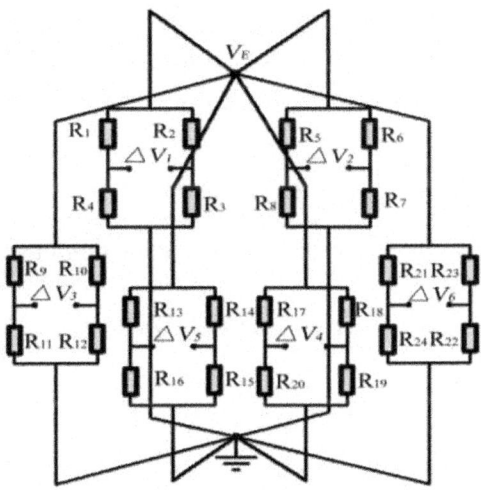

Figure 10: Electrical circuits of the strain gauges.

Subjected to a Cutting Force Fx

While the system is individually applied with the tangential cutting force F_x (similar to F_y), the resistances changes in the way as defined in the Table 4. Therefore, the output signals can be expressed as follows:

$$\Delta V_1 = \frac{V_E}{4}\left(\frac{\Delta R_1}{R_1} - \frac{\Delta R_2}{R_2} + \frac{\Delta R_3}{R_3} - \frac{\Delta R_4}{R_4}\right) = \frac{V_E}{4}\left[2\left(\frac{\Delta R_1}{R_1}\right)_\varepsilon - 2\left(\frac{\Delta R_2}{R_2}\right)_\varepsilon\right] = \frac{V_E}{4}\left[2\left|\left(\frac{\Delta R_1}{R_1}\right)_\varepsilon\right| + 2\left|\left(\frac{\Delta R_2}{R_2}\right)_\varepsilon\right|\right] \tag{21}$$

$$\Delta V_2 = \frac{V_E}{4}\left(\frac{\Delta R_5}{R_5} - \frac{\Delta R_6}{R_6} + \frac{\Delta R_7}{R_7} - \frac{\Delta R_8}{R_8}\right) = 0 \tag{22}$$

$$\Delta V_3 = \frac{V_E}{4}\left(\frac{\Delta R_9}{R_9} - \frac{\Delta R_{10}}{R_{10}} + \frac{\Delta R_{11}}{R_{11}} - \frac{\Delta R_{12}}{R_{12}}\right) = 0 \tag{23}$$

$$\Delta V_4 = \frac{V_E}{4}\left(\frac{\Delta R_{17}}{R_{17}} - \frac{\Delta R_{18}}{R_{18}} + \frac{\Delta R_{19}}{R_{19}} - \frac{\Delta R_{20}}{R_{20}}\right) = 0$$

(24)

$$\Delta V_5 = \frac{V_E}{4}\left(\frac{\Delta R_{13}}{R_{13}} - \frac{\Delta R_{14}}{R_{14}} + \frac{\Delta R_{15}}{R_{15}} - \frac{\Delta R_{16}}{R_{16}}\right) = 0$$

(25)

$$\Delta V_6 = \frac{V_E}{4}\left(\frac{\Delta R_{21}}{R_{21}} - \frac{\Delta R_{22}}{R_{22}} + \frac{\Delta R_{23}}{R_{23}} - \frac{\Delta R_{24}}{R_{24}}\right) = 0$$

(26)

where the $(\Delta R_i/R_i)_\varepsilon$ represents the resistance variation rate of the gauge R_i, caused by the strain change. The above analysis shows that only the Fx-bridge is dependent on the applied cutting force F_x, and the outputs of else bridge circuits are independent.

Under the Normal Cutting Force Fz

Under a cutting force Fz action the corresponding output voltages of the sensor can be written as follows:

$$\Delta V_3 = \frac{V_E}{4}\left(\frac{\Delta R_9}{R_9} - \frac{\Delta R_{10}}{R_{10}} + \frac{\Delta R_{11}}{R_{11}} - \frac{\Delta R_{12}}{R_{12}}\right) = \frac{V_E}{4}\left[2\left(\frac{\Delta R_9}{R_9}\right)_\varepsilon - 2\left(\frac{\Delta R_{10}}{R_{10}}\right)_\varepsilon\right] = \frac{V_E}{4}\left[2\left|\left(\frac{\Delta R_9}{R_9}\right)_\varepsilon\right| + 2\left|\left(\frac{\Delta R_{10}}{R_{10}}\right)_\varepsilon\right|\right]$$

(27)

$$\Delta V_1 = \Delta V_2 = \Delta V_4 = \Delta V_5 = \Delta V_6 = 0$$

(28)

Under the Cutting Moment about the Tangential Axis Mx (Similar to My)

Under the cutting moment Mx action the sensor output signals can be derived as:

$$\Delta V_4 = \frac{V_E}{4}\left(\frac{\Delta R_{17}}{R_{17}} - \frac{\Delta R_{18}}{R_{18}} + \frac{\Delta R_{19}}{R_{19}} - \frac{\Delta R_{20}}{R_{20}}\right) = \frac{V_E}{4}\left[2\left(\frac{\Delta R_{17}}{R_{17}}\right)_\varepsilon - 2\left(\frac{\Delta R_{18}}{R_{18}}\right)_\varepsilon\right] = \frac{V_E}{4}\left[2\left|\left(\frac{\Delta R_{17}}{R_{17}}\right)_\varepsilon\right| + 2\left|\left(\frac{\Delta R_{18}}{R_{18}}\right)_\varepsilon\right|\right]$$

(29)

$$\Delta V_1 = \Delta V_2 = \Delta V_3 = \Delta V_5 = \Delta V_6 = 0$$

(30)

Subjected to a Cutting Moment Mz

Under a cutting moment an Mz action the lamellas with strain gauges will be subjected to a tensile/compressive stress. As a result, the output signals of the sensor can be expressed as:

$$\Delta V_6 = \frac{V_E}{4} \left(\frac{\Delta R_{21}}{R_{21}} - \frac{\Delta R_{23}}{R_{23}} + \frac{\Delta R_{22}}{R_{22}} - \frac{\Delta R_{24}}{R_{24}} \right) - \frac{V_E}{4} \left[2 \left(\frac{\Delta R_{21}}{R_{21}} \right)_\varepsilon - 2 \left(\frac{\Delta R_{23}}{R_{23}} \right)_\varepsilon \right] - \frac{V_E}{4} \left[2 \left| \left(\frac{\Delta R_{21}}{R_{21}} \right)_\varepsilon \right| + 2 \left| \left(\frac{\Delta R_{23}}{R_{23}} \right)_\varepsilon \right| \right]$$

(31)

$$\Delta V_1 = \Delta V_2 = \Delta V_3 = \Delta V_4 = \Delta V_5 = 0$$

(32)

Hence, each force/moment component of the cutting load can only generate an output voltage of the corresponding bridge circuit, which means the coupling among the sensor components is negligible in theory. In addition, it is seen that the electrical circuits characterized with intrinsic compensation for temperature drift.

CALIBRATION AND CHARACTERIZATION OF THE SENSOR

A prototype of the sensor system for cutting force measurement in machining processes with an integrated electric circuit has been constructed, as shown in Figure 11. The sensor has a total size of Φ 80 mm × 42 mm. The measurement range of the system is set to 0–250 N for cutting force F_z, ±200 N for cutting forces in shear directions (F_x and F_y), and ±10 Nm for cutting moments around normal axis, and ±8 Nm for cutting moments around the shear axis.

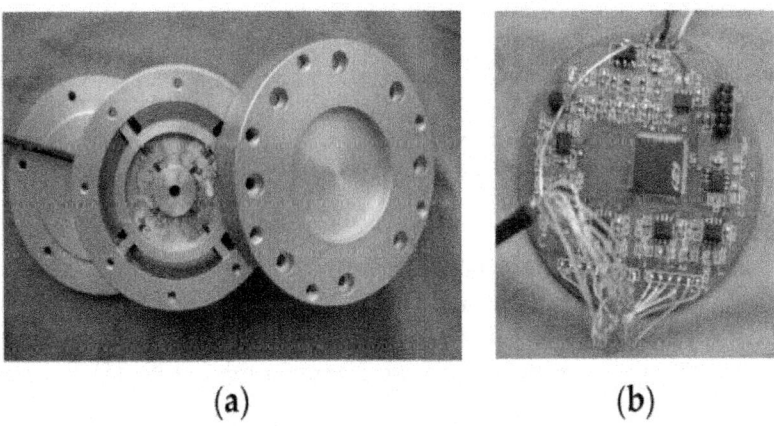

(a)   (b)

Figure 11: Prototype of the sensor system for cutting force measurement: (a) the upper adapter, EE and the lower adapters; (b) the integrated electric circuit.

Precise calibration and decoupling of multi-dimensional cutting force sensor system is critical and must be addressed. The coupling effects among cutting force and moment components are depicted in Figure 12. The relationship between the applied cutting load and output of the system could be expressed as:

$$\Delta V = FW + b \tag{33}$$

With the method and procedure previously proposed in [17], calibration and decoupling experiment of the proposed system are carried out. Table 5 shows the outputs of the sensor under different applied cutting forces and moments. Specifically, the outputs of the components have a good symmetry about the zero point.

Table 5: The outputs of the sensor

Component	F_x	F_y	F_z	M_x	M_y	M_z
Applied load	-	-	-	-	-	-10 Nm
Corresponding output	-	-	-	-	-	-3.40 V
Applied load	-200 N	-200 N	0 N	-8 Nm	-8 Nm	-8 Nm
Corresponding output	-4.48 V	-4.35 V	-0.0034 V	-4.03 V	-3.91 V	-2.72 V
Applied load	-150 N	-150 N	50 N	-6 Nm	-6 Nm	-6 Nm
Corresponding output	-3.41 V	-3.31 V	0.73 V	-3.08 V	-2.98 V	-2.04 V
Applied load	-100 N	-100 N	100 N	-4 Nm	-4 Nm	-4 Nm
Corresponding output	-2.35 V	-2.28 V	1.40 V	-2.12 V	-2.06 V	-1.36 V
Applied load	-50 N	-50 N	150 N	-2 Nm	-2 Nm	-2 Nm
Corresponding output	-1.27 V	-1.25 V	2.07 V	-1.15 V	-1.10 V	-0.68 V
Applied load	0 N	0 N	200 N	0 Nm	0 Nm	0 Nm
Corresponding output	-0.00153 V	-0.00168 V	2.72 V	0.0006 V	0.00229 V	0.0032 V
Applied load	50 N	50 N	250 N	2 Nm	2 Nm	2 Nm
Corresponding output	1.28 V	1.26 V	3.35 V	1.15 V	1.11 V	0.66 V
Applied load	100 N	100 N	-	4 Nm	4 Nm	4 Nm
Corresponding output	2.40 V	2.28 V	-	2.11 V	2.05 V	1.37 V
Applied load	150 N	150 N	-	6 Nm	6 Nm	6 Nm
Corresponding output	3.43 V	3.32 V	-	3.09 V	2.99 V	2.08 V
Applied load	200 N	200 N	-	8 Nm	8 Nm	8 Nm
Corresponding output	4.5 V	4.37 V	-	4.03 V	3.91 V	2.80 V
Applied load	-	-	-	-	-	10 Nm
Corresponding output	-	-	-	-	-	3.50 V

Table 6 shows the performance of the proposed cutting force measurement system. From the table, these values show that the proposed system is superior in the maximum coupling and nonlinearity errors. Besides, the system still suffers from slight coupling due to the monolithic EE structure and the attachment error of the strain gauges.

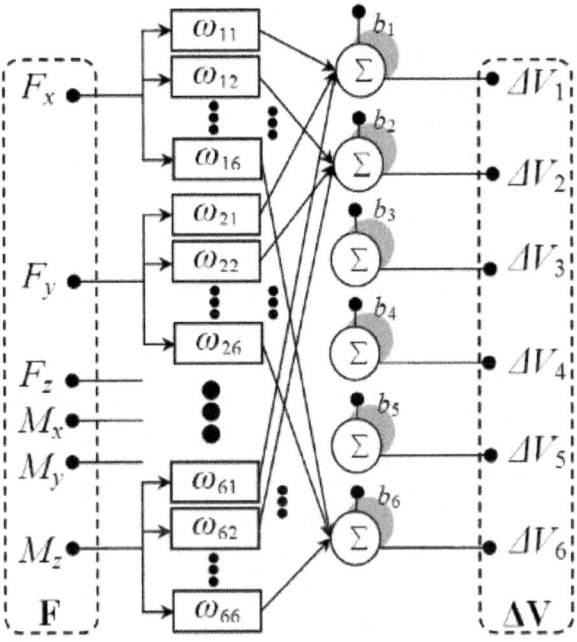

Figure 12: Coupling model of multi-dimensional cutting force sensor system.

Table 6: The performance of the system

Component	Fx	Fy	Fz	Mx	My	Mz
Sensitivity	0.02245 V·N⁻¹	0.02180 V·N⁻¹	0.0134 V·N⁻¹	0.5038 V·(Nm)⁻¹	0.4888 V·(Nm)⁻¹	0.345 V·(Nm)⁻¹
Maximum coupling error	1.07%	1.38%	0.41%	1.47%	1.09%	0.39%
Maximum nonlinearity error	1.73%	1.94%	1.87%	−1.77%	−1.69%	1.15%

CONCLUSIONS

In the present study, a new sensor system for cutting force measurement with a novel decoupling EE has been proposed. The sensor is intended to be used in machining processes for measuring tangential and normal forces F_x, F_y and F_z as well as the moments terms M_x, M_y and M_z simultaneously. By performing a Simulation-Driven Optimization in conjunction with the FEA, the best combination of design parameters is efficiently and scientifically determined. It is revealed that the proposed EE structure with double circular diaphragms may be adopted by other force/moment sensors. As shown in the Table 7, the obtained results demonstrate that the proposed system has satisfied performance and can be utilized to detect static cutting F/M.

Table 7: The performance comparisons with some proven reference sensors

Developer	Approach & Measurement Principle	Size (mm)	No. of Axes	Sensitivity	Maximum Relative Error
Tuysuz, Altintas, Feng [9]	Indirect & prediction model	n.a.	5	n.a.	8.5%
Rao, Gao, Friedrich [13]	Direct & Piezoelectric	Integrated into system	1	7 mV/gm	9.8%
Kim, Kim [15]	Direct & strain gauge and piezo-film accelerometer	$40 \times 70 \times 26$	2	$3\,\text{mV}\cdot\text{N}^{-1}$	n.a.
Yaldız, Ünsaçar [23]	Direct & strain gauge and piezoelectric accelerometer	$100 \times 100 \times 50$	3	$0.1\,\text{mV}\cdot\text{N}^{-1}$	<5%
Liu, Zhou, Tao, Tan [24]	Direct & strain gauge and fiber Bragg grating sensor	n.a.	3	n.a.	6.23%
Our approach	Direct & strain gauge	$\Phi\,80 \times 42$	6	$0.0134\,\text{V}\cdot\text{N}^{-1}$ and $0.345\,\text{V}\,(\text{Nm})^{-1}$	<5%

The sensor system is now being applied to measure cutting loads in machining processes, and reliable results are expected based on the findings reported in this manuscript. For more information about the dynamic cutting F/M measurement and actual cutting experiments, readers should look to our future work with this sensor system.

ACKNOWLEDGMENTS

This work was supported in part by the National Nature Science Foundation of China (NSFC 61203207 and 61573134), the Young Core Instructor Foundation of Hunan Provincial Institutions of Higher Education, the Hunan Provincial Natural Science Foundation of China (14JJ1011), the Changsha Science and Technology Program (k1501009-11), and State Key Laboratory of Advanced Design and Manufacturing for Vehicle Body (31565006).

AUTHOR CONTRIBUTIONS

Qiaokang Liang designed the research, Dan Zhang and Gianmarc Coppola designed the decoupling method. Yaonan Wang and Wei Sun decoupled and characterized the sensors. Jianxu Mao and Yunjian Ge analyzed and interpreted the data. Qiaokang Liang and Gianmarc Coppola co-wrote the manuscript.

REFERENCES

1.   Elbestawi, M.A.; Dumitrescu, M. Tool condition monitoring in machining-neural networks. *Inf. Technol. Balanced Manuf. Syst.* 2006, *220*, 5–16.

2.   Matsushima, K.; Bertok, P.; Sata, T. In process detection of tool breakage by monitoring the spindle current of a machine tool. *Meas. Control Batch Manuf.* 1982, *1*, 145–154.

3. Penedo, F.; Haber, R.E.; Gajate, A. Hybrid incremental modeling based on least squares and Fuzzy-NN for monitoring tool wear in turning processes. *IEEE Trans. Ind. Inform.* 2012, *8*, 811–818.

4. Zhou, J.H.; Pang, C.K.; Lewis, F.L. Intelligent diagnosis and prognosis of tool wear using dominant feature identification. *IEEE Trans. Ind. Inform.* 2009, *5*, 454–464.

5. Constantinides, N.; Bennett, S. An investigation of methods for the on-line estimation of tool wear. *Int. J. Mach. Tool. Manuf.* 1987, *27*, 225–237.

6. Shin, B.C.; Ha, S.J.; Cho, M.W.; Seo, T.I.; Yoon, G.S.; Heo, Y.M. Indirect cutting force measurement in the micro end-milling process based on frequency analysis of sensor signals. *J. Mech. Sci. Technol.* 2010, *24*, 165–168.

7. Kim, T.Y.; Kim, J. Adaptive cutting force control for a machining center by using indirect cutting force measurements.*Int. J. Mach. Tool. Manuf.* 1996, *36*, 925–937.

8. Auchet, S.; Chevrier, P.; Lacour, M.; Lipinski, P. A new method of cutting force measurement based on command voltages of active electro-magnetic bearings. *Int. J. Mach. Tool. Manuf.* 2004, *44*, 1441–1449.

9. Tuysuz, O.; Altintas, Y.; Feng, H.Y. Prediction of cutting forces in three and five-axis ball-end milling with tool indentation effect. *Int. J. Mach. Tool. Manuf.* 2013, *66*, 66–81.

10. Tsai, C.; Liao, Y. Cutting force prediction in ball-end milling with inclined feed by means of geometrical analysis. *Int. J. Adv. Manuf. Technol.* 2010, *46*, 529–541.

11. Fussell, B.K.; Jerard, R.B.; Hemmett, J.G. Modeling of cutting geometry and forces for 5-axis sculptured surface machining. *Comput. Aided Des.* 2003, *35*, 333–346.

12. Ferry, W.B.; Altintas, Y. Virtual five-axis flank milling of jet engine impellers—Part I: Mechanics of five-axis flank milling. *J. Manuf. Sci. Eng.* 2008, *130*.

13. Rao, B.C.; Gao, R.X.; Friedrich, C.R. Integrated force measurement for on-line cutting geometry inspection. *IEEE Trans. Instrum. Meas.* 1995, *44*, 977–980.

14. Kim, J.H.; Chang, H.K.; Han, D.C.; Jang, D.Y.; Oh, S.I. Cutting force estimation by measuring spindle displacement in milling process. *CIRP Ann-Manuf. Technol.* 2005, *54*, 67–70.

15. Kim, J.D.; Kim, D.S. Development of a combined-type tool dynamometer with a piezo-film accelerometer for an ultra-precision lathe. *J. Mater. Process. Technol.* 1997, *71*, 360–366.

16. Jin, W.L.; Venuvinod, P.K.; Wang, X. An optical fibre sensor based cutting force measuring device. *Int. J. Mach. Tools Manuf.* 1995, *35*, 877–883.

17. Liang, Q.; Zhang, D.; Ge, Y.; Song, Q. A novel miniature four-dimensional force/torque sensor with overload protection mechanism. *IEEE Sens. J.* 2009, *9*, 1741–1747.

18. Liang, Q.; Zhang, D.; Coppola, G.; Wang, Y.; Sun, W.; Ge, Y. Multi-Dimensional MEMS/Micro Sensor for Force and Moment Sensing: A Review. *IEEE Sens. J.* 2014, *14*, 2643–2657.

19. Park, S.S. High Frequency Bandwidth Cutting Force Measurements in Milling Using the Spindle Integrated Force Sensor System. Ph.D. Thesis, University of British Columbia, Vancouver, BC, Canada, 2003.

20. Jantunen, E. A summary of methods applied to tool condition monitoring in drilling. *Int. J. Mach. Tools Manuf.* 2002, *42*, 997–1010.

21. Bayo, E.; Stubbe, J.R. Six-axis force sensor evaluation and a new type of optimal frame truss design for robotic applications. *J. Robot. Syst.* 1989, *6*, 191–208.

22. Svinin, M.M.; Uchiyama, M. Optimal geometric structures of force/torque sensors. *Int. J. Robot. Res.* 1995, *14*, 560–573.

23. Yaldız, S.; Ünsaçar, F. A dynamometer design for measurement the cutting forces on turning. *Measurement* 2006, *39*, 80–89.

24. Liu, M.; Zhou, Z.; Tao, X.; Tan, Y. A dynamometer design and analysis for measurement the cutting forces on turning based on optical fiber Bragg Grating sensor. In Proceedings of the 2012 10th World Congress on Intelligent Control and Automation (WCICA), Beijing, China, 6–8 July 2012.

Chapter 4

SELECTION OF CUTTING INSERTS IN DRY MACHINING FOR REDUCING ENERGY CONSUMPTION AND CO$_2$ EMISSIONS

Rosario Domingo[1], Marta María Marín[1], Juan Claver[1], and Roque Calvo[2]

[1]Departamento de Ingeniería de Construcción y Fabricación, Universidad Nacional de Educación a Distancia (UNED), C/Juan del Rosal 12, Madrid 28040, Spain

[2]Departamento de Ingeniería Mecánica, Química y Diseño Industrial, Universidad Politécnica de Madrid, Ronda de Valencia 3, Madrid 28012, Spain

ABSTRACT

Manufacturing processes are responsible for climate change due to the emissions produced as result of energy consumption. This paper analyzes the influence of the cutting conditions and the characteristics of cutting tools on the energy required in machining processes and the carbon dioxide equivalent (CO$_2$-eq) emissions generated per material removed ratio (MRR) in an effort to define common criteria for using cutting inserts in a sustainable manner. Consequently, four cutting inserts were evaluated during the turning of Ti6Al4V alloy. An experimental and statistical methodology that combined the orthogonal array L36, the signal-to-noise ratio under the "small is better" criterion of Taguchi, and a multifactor analysis of variance was used. The effects of the geometry, material and coating of the tool and the cutting conditions on the energy and the carbon footprint during the manufacturing process were analyzed. The results show that a high tool cutting length and a high cutting depth are significant common factors, whereas the coating-cutting depth and cutting length-cutting speed are significant common interactions for both the energy/MRR ratio and the CO$_2$-eq emissions/MRR ratio, and the coating-cutting speed exhibits a significant interaction for emissions. The outcomes show that the lifespan of the tool has little influence on the total emissions, at the time that the methodology is able to identify the most appropriate manner to calculate energy.

INTRODUCTION

Intense industrial activity and manufacturing processes require high energy consumption and thereby the generation of greenhouse gas (GHG) emissions, which have negative consequences for the preservation of resources and the environment, due to their contribution to global warming. These GHG emissions include carbon dioxide (CO_2), the main contaminant gas generated in the world, and other gases such as methane, nitrous oxide and chlorofluorocarbons which can be measured in units of CO_2 equivalents (CO_2-eq) [1]. In the manufacturing field, designs that reduce energy use and emissions are the main issues considered by companies in their attempt to move toward greater sustainability [2]. Energy efficiency is a current concern, and policies regarding it affect the environment because this efficiency can contribute to reducing emissions and favor decreasing the greenhouse effect [3]. Some companies have their own policies and focus their production on clean manufacturing [4] and models are being developed to minimize environmental impacts while maximizing resource efficiency in particular cases, such as gear manufacturing [5]. In the industrial field, where innovation is continuous, the influence of many factors (e.g., equipment, materials and tools) must be considered. New materials and new processes or modifications of old processes appear constantly in the search for high performance due to global competition.

Although processes are often studied by considering their energy consumption [6], it is convenient to investigate the GHG emissions generated by particular industrial activities. Problems attributed to CO_2 emissions have been exposed in some products [7] due to their contribution to climate change. Recently, criticism has arisen regarding the suitability of GHG emission reductions due to their weak economic impacts [8], but some researchers have demonstrated that emissions reductions can be compatible with costs savings in a manufacturing plant through environmental innovation [9] and that the environmental commitment can improve the industrial plant's productivity [10]. These items are also inherent in mechanical manufacturing processes, in which machining is commonly used due to its versatility. Thus, in these processes, features such as machines, fixtures, tools and cutting fluids, including lubricants and coolants, should be studied because these affect the results of the operations performed. The equipment used in these processes, namely, machine tools, must be designed in a more efficient manner to survive in the market [11], but the current focus is on the process. For these reasons, machining processes that require material to be removed from an initial part must be thoroughly analyzed. These processes require greater power and higher energy causing high CO_2-eq emissions, particularly for harder materials, where higher cutting speed is required. High performance also requires a high feed rate and high cutting depth because these allow faster material removal. For these reasons, some energy consumption models have been proposed [12], and different strategies for the design of tool paths have been studied [13] with the objective of consuming less energy in the process. Many efforts have been devoted to the study of cutting fluids and processes in dry [14] or cryogenic [15,16] conditions, and these have yielded better results compared with wet machining. The tool is an important element because it is responsible for material removal; thus, tool design, including the geometry and the materials used, has been a priority in this field [17]. Consequently, tool design has evolved in recent decades but continues to focus on achieving higher performance without considering the sustainability of the process.

The titanium alloy Ti6Al4V is commonly used in the aeronautical and biomedical industries, among others, due to its good mechanical proprieties and its biocompatibility in medical uses. In the machining process, turning operations are utilized to obtain the final geometry of many components. The difficulty of machining is a common topic in the literature [18,19]. This alloy is very hard, and it is susceptible to temperature increments during machining, which can induce a phase change. Moreover, the thermal conductivity of this alloy is low compared to that of other materials, such as steel, which causes an increase in temperature in the machining zone. In general, titanium consumes more energy and produces more emissions than other structural materials, such as stainless steel; thus, its convenience of use must take into account the product lifespan [20]. This titanium alloy has been selected for this study due to its widespread use and the difficulties associated with its machining. In particular, for roughing operation, it is important from a sustainable perspective to determine the insert and the cutting conditions that would result in low energy consumption and CO_2-eq emissions.

Dry machining is considered a more sustainable process than machining with cutting fluids due to the absence of lubricants and coolants, which results in a reduced use of resources [14,21]. Moreover, in the dry turning of titanium alloy, the cutting forces required are lower than those required using cutting fluids [18], and a better surface quality is achieved [19] than in wet machining. In this sense, the selection of a dry process could result in a reduction of energy consumption. However, dry machining results in higher wear of the cutting inserts [19], which can be counterweighted with the use of coating inserts to increase tool life [22].

Some strategies and models have been developed mainly to minimize the energy consumption in machining. Rajemi et al.[23] proposed a model that looks for the selection of the optimal turning conditions based on minimum energy considerations. Balogun and Mativenga [24] then improved the model to introduce direct energy requirements, and Peng *et al.* [25] developed an energy model implementation for future CNC machining systems based on function blocks. Iqbal *et al.*[26] established a set of rules utilizing fuzzy logic and found that a long tool life is associated with more energy consumption and low productivity. In addition, Schultheiss *et al.* [27] showed that it is possible to increase tool life through the use of worn tools in secondary operations. Mativenga and Rajemi [28] showed that the selection of cutting conditions based on a minimum energy footprint criterion can lead to an important decrease in the energy footprint of the machining process compared with that obtained the cutting conditions recommended by tool manufactures; Fang *et al.* [29] applied a model of general multi-objective mixed-integer linear programming to machining in an attempt to reduce power consumption and the carbon footprint. Finally, Yingjie [30] concluded that it is necessary to research energy efficiency techniques in machining processes through the optimization of mechanical configurations and cutting parameters, whereas that Peng and Xu [31] affirmed that a future comprehensive data analysis in terms of energy consumption should include materials, machine tools, cutting tools and energy. Despite these advances, it is necessary to know which cutting conditions are significant in terms of their effects on energy or CO_2-eq emissions. Moreover, conditions can be easily comparable if both the energy and emissions are related to the material removed ratio (MRR).

Although many problems are related to policies, particularly those linked to the type of energy generated, it can be made decisions on industrial processes to mitigate the effects. The objectives of this paper are:(i) to evaluate the influence of the cutting conditions and insert characteristics on the energy required and the CO_2-eq emissions per MRR during the dry machining of Ti6Al4V alloy, with the aim of defining parameters that contribute to reducing the influence of this process on greenhouse gas emissions, and (ii) to determine the contribution of cutting tools to CO_2-eq emissions.

EXPERIMENTAL SECTION

This section describes the experimental procedure used to obtain the values of the forces that allow for the calculation of the energy, the calculation procedure to consider the energy mix and the emissions factor of materials, as well as the statistical methodology followed according to Taguchi's method.

Experimental Procedure

Ti6Al4V alloy bars have been used due to the importance of this material in the industry. The operation selected is turning, particularly orthogonal cutting, because cylindrical parts are components of many titanium products, such as biomedical implants, vessels, hubs, fasteners, and many other components. Some main properties of this alloy are shown inTable 1 [32].

Table 1: Proprieties of the Ti6Al4V alloy

Density	4.43 g/cm^3
Hardness, Brinell	360
Thermal conductivity	6.70 W/mK
Melting point	1604–1660 °C
Beta transus	980 °C

Several high-performance inserts have been chosen; they are recommended by the tool manufacturers for the machining of titanium, particularly SECO (Seco Group, Fagersta, Sweden), the cutting tool manufacturer selected in these tests. According ISO 1832 nomenclature [33], the inserts are as follows: CNMG 120408 MF1 890 (hereinafter CNMG), CNMG 120408 MF1 CP500 (hereinafter CNMG PVD), WNMG 060408 MF1 890 (hereinafter WNMG) and WNMG 060408 MF1 CP500 (hereinafter WNMG PVD). Thus, uncoated and coated inserts are considered. The coating is achieved by physical vapor deposition (PVD). The material coded 890 is cobalt-cemented tungsten carbide, WC with 6% Co as binder, manufactured by powder metallurgy technology. These coatings are more environmentally friendly than others, such as those generated by chemical vapor deposition (CVD), which requires more operations for their manufacturing [34]; this feature was taken into account when selecting the coatings investigated in this study. These tools have been selected because they provide good results in Ti6Al4V due to good machinability [35]. The tool chip breaker (MF1) used is a medium type that is adequate for finishing and roughing operations is shown in Table 2 on the images of each insert. The difference between CNMG and WNMG is the number of edges (four and six, respectively) and the cutting length of the edge (*le*) is 12.7 and 9.525 mm, respectively. According to the ISO nomenclature, all of the inserts have the same cutting angle (80°), clearance angle (0°), clearance thickness (0.13 mm), and nose radius (0.8 mm). The weight of the insert type CNMG is 0.02223 kg, and the weight of insert type WNMG is 0.01075 kg. The coating is titanium nitride (TiN) plus titanium aluminum nitride (TiAlN) via PVD. This coating is selected because its good wear resistance, increasing the tool life [36].

Table 2: Characteristics of the inserts

Code	WNMG	WNMG PVD	CNMG	CNMG PVD
Insert with pointed tool chip breaker				
Material	WC	WC	WC	WC
Coating	-	TiN + TiAlN	-	TiN + TiAlN
Number of edges	6	6	4	4
Cutting length, le (mm)	9.525	9.525	12.7	12.7

The ranges of cutting conditions are the following: cutting depths (d), 1, 2 and 3 mm, feed rate (f), 0.1 and 0.2 mm/rev; and cutting speed (vc), 50, 100 and 150 m/min. Under these cutting conditions, it is possible to find regions of low, moderate and high tool wear [37]. To maintain a constant spindle speed, bars of different diameters were used: 17.2 mm for vc = 50 m/min, 34.4 mm for vc = 100 m/min and 51.6 mm for vc = 150 m/min. The tests were performed in dry conditions because it is more environmental friendly to avoid the use of cutting fluids [14].

The equipment used includes a Kistler-type 9257B piezoelectric dynamometer with a Kistler-type 5070A multi-channel charge amplifier (Kistler Group, Winterthur, Switzerland) installed in an EMCO CNC lathe (EMCO Maier GesmbH, Hallein, Austria). The data acquisition software to convert the electrical signal into mechanical units is DasyLab (Measurement Computing Corporation, Norton, MA, USA). The data collected directly include the cutting force (F_c), back force (F_p) and feed force (F_f) (see Figure 1). In this manner, all of the forces are considered for a total energy calculation. The energy required for the tool change was also considered. These aspects are important because in machining, the total energy is much higher than the cutting energy [38]. With this information, it is possible to calculate the cutting, feed, active and drive power. From these values, the ratio of energy to MRR can be finally calculated. The machine efficiency is assumed to be 90%, which is within the typical machine efficiency range of 80% to 90% [17]. This efficiency contributes to the absolute value of the energy, but considering the same technology level of the machines, it does not contribute to a different result about insert tools behavior.

The tool life was analyzed by a three-dimensional measurement device with a TESA VISIO laser sensor (TESA SA, Renens, Switzerland), which can determine the wear of each cutting edge at each tool insert. Although, the energy consumption can increase due to tool wear, because the forces required are higher [39], in these experiments, each insert is used for its lifespan.

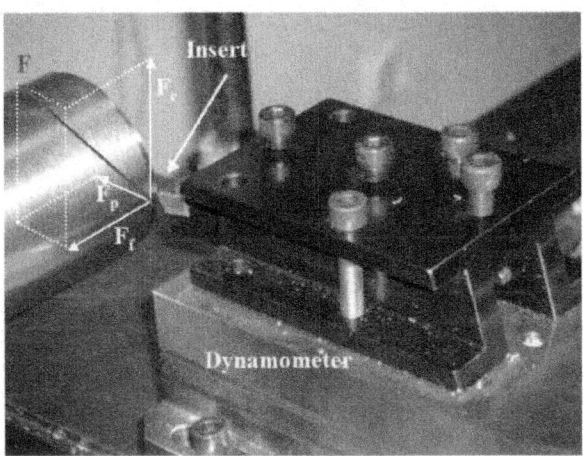

Figure 1: Assembly of the dynamometer, insert and part, and types of forces.

Calculation Procedure

This calculation considers the emissions derived from the energy consumed during the machining and the manufacturing of the inserts. The emissions resulting from the titanium manufacturing are not considered because they are common in all cases; thus, they do not contribute to identifying the ideal conditions. CO_2-eq emissions are calculated considering the Spanish energy mix. Table 3 describes the Spanish energy mix; as shown in the table, the emission factor is 0.44 tonnes CO_2-eq/GJ [40]. In this country, there is a high percentage of clean energy, so the results could be considerably higher in other countries. Moreover, the emissions derived from the inserts, particularly from WC manufactured by a sintering process, have a coefficient of 1.34 tonnes of CO_2-eq per tonne of manufactured product [41], in addition to the manufacturing process of the insert coatings, which are formed by physical vapor deposition, which adds 0.08 tonnes of CO_2-eq per tonne of manufactured product [41]. Because these coefficients are lifecycle emission contributions, all of these coefficients and the weight of the inserts were taken into account in the calculation of CO_2-eq emissions, which were performed using SimaPro software (Pré Consultants, Amersfoort, The Netherlands) [42], to obtain the results presented in Section 3.

Table 3: Origin of electrical energy

Origin	Percentage (%)
Coal	11.3
Combined cycle	9.1
Wind	25.0
Hydro	16.2
Nuclear	20.5
Photovoltaic solar	2.4
Renewable thermal	2.1
Thermal solar	0.8
Cogeneration and remaining	12.6

In these calculations, it is necessary to consider the number of inserts required for machining Ti6Al4V alloy. In this sense, the tool life is an important variable. ISO 3685 [43] establishes the limit on flank tool wear (VB) at 0.3 mm, which has been considered in the calculation of the number of tools required during the process and will affect the emissions outcomes due to the production of tools. Once an insert achieves a wear of 0.3 mm, a new insert replaces it. Preliminary tests were executed to determine the tool life, measuring the growth of flank wear, and the machining time for each edge and each cutting condition. Once the edge life is known, the tool was changed during the tests at the appropriate time, and the CO_2-eq emissions generated from the energy consumed during the times of change were taken into account. The overall operation time was the sum of insert lifespan plus an average of 90 s of change time per insert. In order to establish proper tool life comparisons, the total number of inserts required to machining a cubic meter of raw material was considered. For energy calculation only machining and change times were considered, but for CO_2-eq calculation tool manufacturing impact was added. This approach allows the tool comparison and also it determines whether the energy required for the process is sufficient for insert selection. Both the energy requirements and the emissions that are common to any type of tool they have not been taken into account, so a differential comparison and selection procedure is provided.

Statistical Procedure

A design of experiments by the Taguchi methodology has been performed [44]. To reduce the number of tests, the orthogonal array L36 ($2^3 \times 3^{13}$) has been chosen because it combines a maximum of three factors with two levels and thirteen factors with three levels. The factors and levels are shown in Table 4, in which the coating, le and f have two levels and vc and d have three levels.

Table 4: Factors and levels of the design of experiments

Factor / Level	Coating	Tool Cutting Length, le (mm)	Feed Rate, f (mm/rev)	Cutting Speed, vc (m/min)	Cutting Depth, d (mm)
1	Uncoated	9.525	0.1	50	1
2	PVD	12.7	0.2	100	2
3	-	-	-	150	3

The measurement is given via the signal-to-noise ratio (S/N), which represents the relationship between the signal or mean and the noise or variance [44]. Taguchi considered several models of S/N: "smallest is better", "nominal is better" and "larger is better"; in this study, "smaller is better" is analyzed because the variables are energy and CO_2-eq emissions. This methodology has been used in the analysis of manufacturing processes [45,46] because it can be used to optimize the parameters that have a major influence on the variables considered. Equation (1) represents the relationship "smaller is better", where n is the number of tests and y_i is the measurement in the ith trial of the objective variable. In this investigation, y_i is the energy and the CO_2-eq emissions by MRR. MRR is the product of the cutting conditions: f, vc and d. For the S/N ratio, "smaller is better" is interpreted as higher values being preferred [44].

$$S/N = -\log\left(\frac{1}{n}\sum_{i=1}^{n} y_i^2\right)$$
(1)

An analysis of variance (ANOVA) has been performed to find the significant factors and interactions in the S/N ratio under a confidence interval of 95%. Therefore, the factor or the interaction is significant if the P-value is less than 0.05 [47]. Moreover, a regression analysis is conducted to verify the relationship between the factors and variables and between energy and emissions. The Statgraphics software (Statpoint Technologies, Inc., Warrenton, VA, USA) was used for the statistical analyses [48].

RESULTS AND DISCUSSION

The factors and levels presented in Table 4 are displayed in greater detail in Table 5 to allow a clear identification of the values associated with each factor at each level. The following results show the value of the total energy per MRR, including the energy achieved by the cutting, back and feed forces.

Table 5: Values of the factors at each level

No. Test	Factors and Levels					Values of Factors at Each Level				
	Tool Characteristics		Cutting Parameters			Tool Characteristics		Cutting Parameters		
	Coating	Tool Cutting Length, le	Feed Rate, f	Cutting Speed, vc	Cutting Depth, d	Coating	Tool Cutting Length, le (mm)	Feed Rate, f (mm/rev)	Cutting Speed, vc (m/min)	Cutting Depth, d (mm)
1	1	1	1	1	1	Uncoated	9.525	0.1	50	1
2	1	1	1	1	2	Uncoated	9.525	0.1	50	2
3	1	1	1	1	3	Uncoated	9.525	0.1	50	3
4	1	2	2	1	1	Uncoated	12.7	0.2	50	1
5	1	2	2	1	2	Uncoated	12.7	0.2	50	2
6	1	2	2	1	3	Uncoated	12.7	0.2	50	3
7	2	1	2	1	1	PVD	9.525	0.2	50	1
8	2	1	2	1	2	PVD	9.525	0.2	50	2
9	2	1	2	1	3	PVD	9.525	0.2	50	3
10	2	2	1	1	1	PVD	12.7	0.1	50	1
11	2	2	1	1	2	PVD	12.7	0.1	50	2
12	2	2	1	1	3	PVD	12.7	0.1	50	3
13	1	1	1	2	1	Uncoated	9.525	0.1	100	1
14	1	1	1	2	2	Uncoated	9.525	0.1	100	2
15	1	1	1	2	3	Uncoated	9.525	0.1	100	3
16	1	2	2	2	1	Uncoated	12.7	0.2	100	1
17	1	2	2	2	2	Uncoated	12.7	0.2	100	2
18	1	2	2	2	3	Uncoated	12.7	0.2	100	3
19	2	1	2	2	1	PVD	9.525	0.2	100	1
20	2	1	2	2	2	PVD	9.525	0.2	100	2
21	2	1	2	2	3	PVD	9.525	0.2	100	3
22	2	2	1	2	1	PVD	12.7	0.1	100	1
23	2	2	1	2	2	PVD	12.7	0.1	100	2
24	2	2	1	2	3	PVD	12.7	0.1	100	3
25	1	1	1	3	1	Uncoated	9.525	0.1	150	1
26	1	1	1	3	2	Uncoated	9.525	0.1	150	2
27	1	1	1	3	3	Uncoated	9.525	0.1	150	3
28	1	2	2	3	1	Uncoated	12.7	0.2	150	1
29	1	2	2	3	2	Uncoated	12.7	0.2	150	2
30	1	2	2	3	3	Uncoated	12.7	0.2	150	3
31	2	1	2	3	1	PVD	9.525	0.2	150	1
32	2	1	2	3	2	PVD	9.525	0.2	150	2
33	2	1	2	3	3	PVD	9.525	0.2	150	3
34	2	2	1	3	1	PVD	12.7	0.1	150	1
35	2	2	1	3	2	PVD	12.7	0.1	150	2
36	2	2	1	3	3	PVD	12.7	0.1	150	3

They are high (see Table 6) compared with those obtained with other materials such as aluminum alloys [49], but are calculated per cubic meter, and the density of Ti6Al4V is higher (see Table 1) that that of aluminum based alloys. Therefore, the required high forces result in high energy consumption and CO_2-eq emissions. The S/N ratios are shown in Table 6. The results show that completely different outcomes are obtained with different cutting conditions. In a first approach, better results for energy and emissions (see line 6 of Table 6) are obtained with an uncoated insert, high cutting length, and the cutting conditions $f = 0.2$ mm/rev, $vc = 50$ m/min and $d = 3$ mm. The worst results are achieved with a coated insert, low cutting length, and cutting conditions $f = 0.2$ mm/rev, $vc = 150$ m/min and $d = 2$ mm (see line 32 of Table 6).

Table 6: Calculated energy/MRR, CO_2-eq emissions and edge life

No. Test	Energy/MRR $(GJ/m^3)^*$	S/N (Energy/MRR)	CO_2-eq Emissions/MRR $(kgCO_2/m^3)^*$	S/N (CO_2-eq Emissions/MRR)	Edge Life (min)
1	1.90	−5.593	869.32	−58.785	14.41
2	1.84	−5.315	831.75	−58.400	10.84
3	1.51	−3.595	683.23	−56.691	8.50
4	1.74	−4.796	803.02	−58.094	19.90
5	0.98	0.127	454.89	−53.158	15.72
6	0.71	2.912	331.79	−50.419	12.80
7	2.35	−7.421	1042.54	−60.362	29.80
8	1.62	−4.179	718.17	−57.125	23.70
9	1.94	−5.774	857.88	−58.669	19.80
10	1.65	−4.330	784.39	−57.891	27.03
11	1.60	−4.100	743.01	−57.420	20.23
12	1.41	−2.963	650.93	−56.271	17.23
13	1.97	−5.891	908.12	−59.163	5.81
14	1.65	−4.372	757.76	−57.591	3.78
15	1.73	−4.746	799.31	−58.054	2.10
16	1.96	−5.864	940.30	−59.465	4.78
17	1.50	−3.551	726.49	−57.225	2.80
18	1.16	−1.303	579.35	−55.259	1.80
19	1.24	−1.899	565.86	−55.054	6.28
20	1.74	−4.800	781.83	−57.862	3.92
21	1.42	−3.069	642.11	−56.152	2.45
22	1.50	−3.515	728.62	−57.250	11.50
23	1.53	−3.714	727.25	−57.234	7.30
24	1.82	−5.187	864.03	−58.731	4.16
25	2.06	−6.291	1001.11	−60.010	1.69
26	1.45	−3.212	681.03	−56.663	1.86
27	1.71	−4.685	816.68	−58.241	0.83
28	1.67	−4.434	863.42	−58.724	1.93
29	1.48	−3.416	737.39	−57.354	1.44
30	1.10	−0.797	587.43	−55.379	0.80
31	1.92	−5.680	883.00	−58.919	2.22
32	2.33	−7.342	1050.74	−60.430	1.66
33	1.54	−3.727	703.30	−56.943	1.10
34	1.34	−2.563	690.78	−56.787	5.20
35	1.40	−2.915	687.68	−56.748	3.67
36	1.41	−3.014	743.04	−57.420	1.43

*These values represent the means of three trials in each test.

As observed in the same table, the edge life is 12.85 min in the best situation (CNMG tool) and 1.72 in the worst case (WNMG PVD tool). Because the edge life varies from 29.8 min (WNMG PVD tool) to 0.8 min (CNMG tool), the influence of tools is not clear. In addition to these conditions, it is important to determine the significant factors, which can provide information about the general behavior.

Energy/MRR

The results of the S/N ratio related to energy/MRR have been subjected to a multifactor ANOVA considering the main factors and their interactions. The outcomes are shown in Table 7. The ANOVA table shows the sum of squares, degrees of freedom, mean square, F-ratio according to a Fisher-Snedecor test and P-value of the main factors (coating, le, f, vc and d) and the interactions (coating-d, and le-vc). Only the significant interactions are given in this table; therefore, the interactions coating-vc, coating-f, coating-le, le-f, le-d, f-d, f-vc and vc-d are not significant. The ANOVA shows two significant factors, le and d, as their P-values are less than 0.05 (see Table 7).

Table 7: ANOVA for the S/N ratio (EnergMRR).

Source	Sum of Squares	Degrees of Freedom	Mean Square	F-Ratio	p-value
Coating	3.589	1	3.589	2.40	0.1470
Le	32.431	1	32.431	21.72	0.0006
f	3.353	1	3.353	2.25	0.1598
Vc	0.490	2	0.245	0.16	0.8505
d	20.780	2	10.390	6.96	0.0099
coating-d	16.560	2	8.280	5.55	0.0197
le-vc	12.886	2	6.443	4.32	0.0387
Residual	17.917	12	1.493		
Total (corrected)	143.98	35			

The percentage of influence is 22.52% le and 14.43% d, obtained by the ratio between the sum of squares of a factor and the total (Table 7). The result for le is interesting because it can influence tool design. Figure 2 shows that an increase in le is recommended. Thus, an increase in le results in an increase in the S/N ratio (Energy/MRR). A possible cause is that the shape of the insert makes chip evacuation difficult and can therefore influence deformed chips and removed chips.

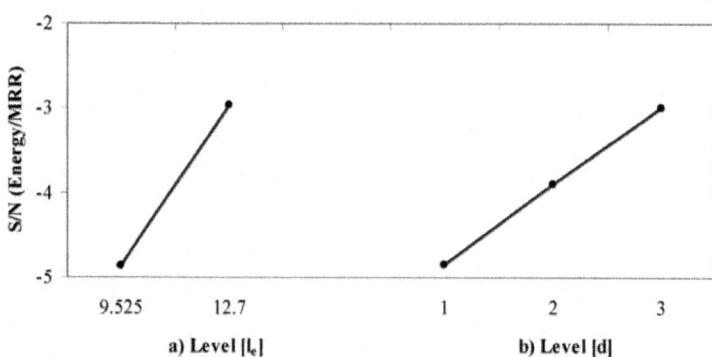

Figure 2.: Values of the S/N ratios of significant main factors for energy/MRR.

Thus, from the energy perspective, CNMG inserts have better results than WNMG inserts. The outcomes of d are also of interest because an increase in d results in a higher S/N ratio. This finding is consistent with the effect of d on the life of the tool, as noted in Table 6; increasing d also increases the tool life. A greater d, 3 mm in the range used, results in more material removed per unit of time; therefore, the high performance is more sustainable.

With respect to significant interactions (see Table 7), only coating-*d* appears and has a positive effect compared to the uncoated insert and a *d* of 3 mm (see Figure 3). This finding is consistent with the results shown in Figure 3; the coating does not positively influence the energy because of the increase in friction.

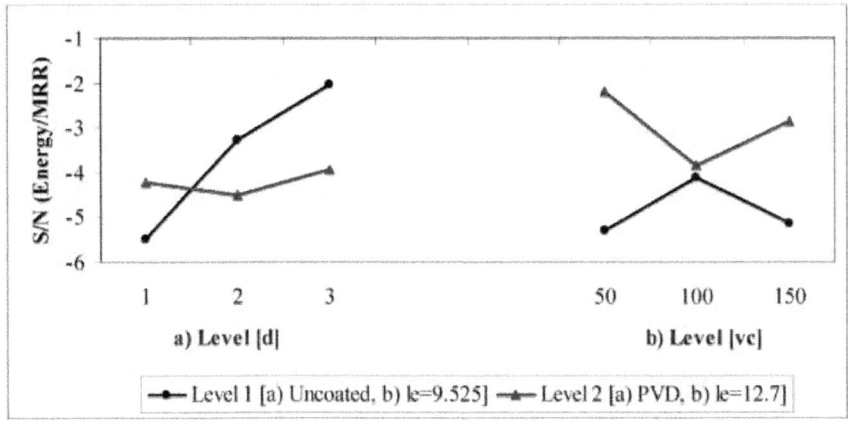

Figure 3:. Values of the S/N ratio of the significant interactions for energy/MRR.

The results for the interaction *le-vc* are different from the results for *le*. An increase in *le* is recommended over a lower *vc*. The outcomes are consistent with the results in Figure 2. Additionally, a lower *vc* (50 mm/min) results in a lower energy consumpt

CO_2-eq Emissions/MRR

Table 8 shows the ANOVA; the factors *le* and *d* are statistically significant, as in the S/N ratio (energy/MRR) and with the same effect (see Figure 4). The significant interactions are coating-*d* and *le-vc*, as for the S/N ratio (energy/MRR), but in this case, coating-*vc* also has a P-value less than 0.05. In this case, the percentage of influence is 16.03% *d* and 12.55% *le*. The influence of *le* is considerably lower than in the study of energy, possibly due to the influence of the coating-*vc* interaction, which is absent in the significant interactions of the S/N ratio (energy/MRR). Thus, although there is a strong relationship between energy and CO_2-eq emissions, in the last case, the influence of tool life can be considered. In fact, the emissions are increased with the number of tools manufactured. The effect of significant interactions is shown in Figure 5.

Table 8:. ANOVA of the S/N ratio (CO_2-eq emissions/R)

Source	Sum of Squares	Degrees of Freedom	Mean Square	F-Ratio	p-value
Coating	2.049	1	2.049	1.38	0.2624
le	16.382	1	16.382	11.05	0.0061
f	4.519	1	4.519	3.05	0.10635
vc	4.468	2	2.234	1.51	0.2606
d	20.934	2	10.467	7.06	0.0094
coating-vc	12.062	2	6.031	4.07	0.0448
coating-d	15.270	2	7.635	5.15	0.0243
le-vc	13.662	2	6.831	4.61	0.0327
Residual	17.784	12	1.482		
Total (corrected)	130.553	35			

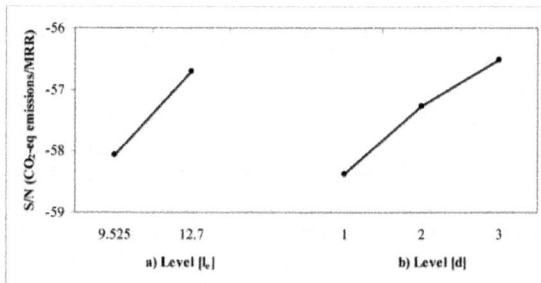

Figure 4.: Values of the S/N ratios of the significant main factors for CO_2-eq emissions/MRR.

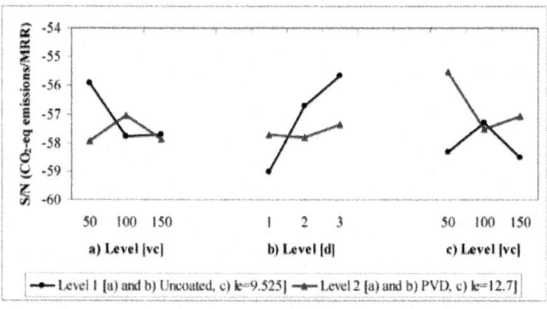

Figure 5:. Values of the S/N ratios of the significant interactions for CO_2-eq emissions/

Contribution of Cutting Inserts to CO_2-eq Emissions

In roughing operations, which are common in machining, coatings do not have a positive influence on energy. Of the inserts evaluated, the most interesting with respect to energy/MRR was CNMG. These last considerations involve sustainability and performance objectives, which balance is important in the industry. Therefore, a longer cutting length has a positive effect. Moreover, this insert has fewer edges than WNMG.

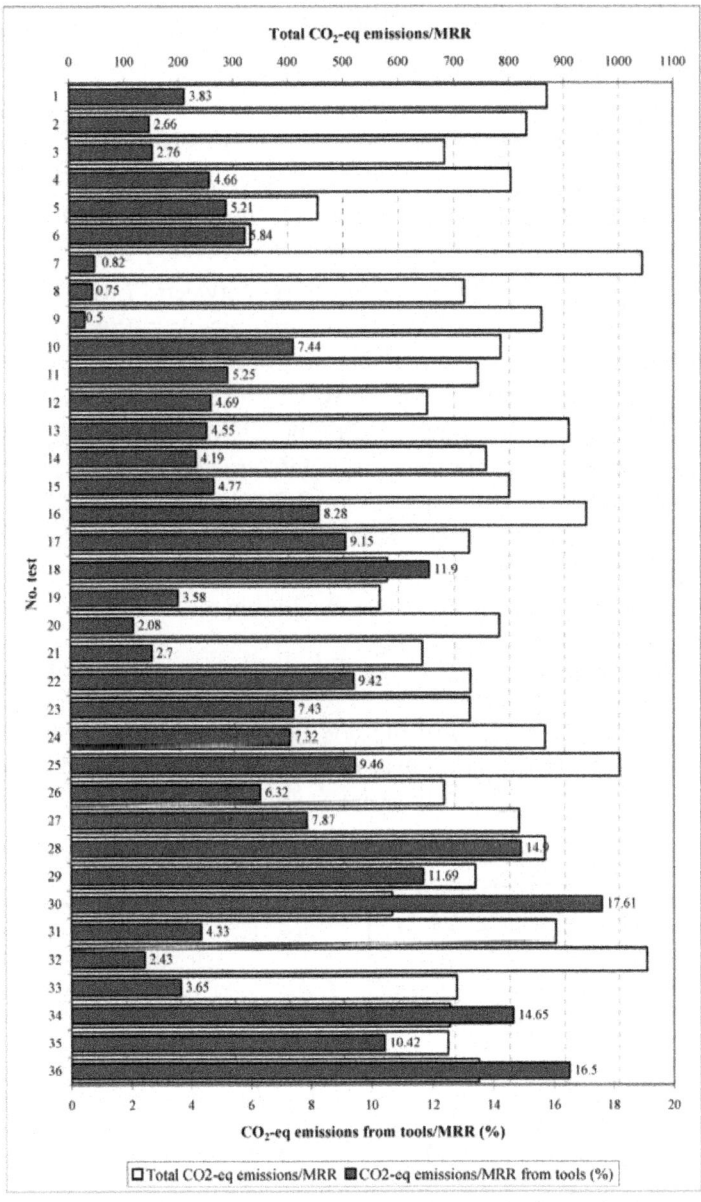

Figure 6:. Contribution of tools to CO_2-eq emissions/MRR as a percentage.

The emissions results for the optimal energy consumption for each insert and the percentages attributed to tools are shown in Figure 6. In this figure, the number of tests is equivalent to the number of tests presented in Table 5 and Table 6 to allow identification of the tool characteristics and cutting parameters. The results show that energy is a more important factor in CO_2-eq emissions, as demonstrated by the finding that higher emissions result from higher energy and not from the emissions associated with tools. The ranking of the inserts from those that generate less emissions to those that generate higher emission is the following: CNMG, CNMG PVD, WNMG PVD and WNMG. These results are consistent with those obtained for energy consumption. As expected, the coating contribution is sufficiently small, and as a result, the percentage is conditioned by the total emissions and not by the coating.

As shown in Figure 6, the percentages are between 17.61% and 0.5%. The uncoated insert type CNMG, with $f = 0.2$ mm/rev, $vc = 150$ m/min and $d = 3$ mm, has the greatest contribution (see line 30 of Table 6), followed by coated insert type CNMG, with $f = 0.1$ mm/rev, $vc = 150$ m/min and $d = 3$ mm (line 36 of Table 6) and the uncoated insert type CNMG, with $f = 0.2$ mm/rev, $vc = 150$ m/min and $d = 50$ mm (line 28 of Table 6). The lowest contributions of the tools are found for coated insert type WNMG, with $f = 0.2$, $vc = 50$ m/min and $d = 1, 2,$ and 3 mm (lines 7, 8 and 9 of Table 6). The lower contribution obtained for the WNMG insert is a consequence of the material, even when these inserts consume more energy in machining te

Evolution of Factors

Figure 7 shows the variation of the S/N ratios as d and le are increased, and Figure 8 shows the same variation as *coating* and le are modified under optimal conditions. Both ratios appear to show a similar evolution.

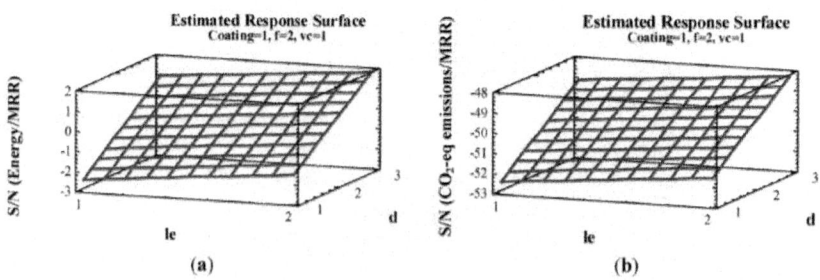

Figure 7:. (a) Values of the S/N ratio (Energy/MRR) with respect to le and d; (b) Values of the S/N ratio (CO_2-eq emissions/MRR) with respect to le and d.

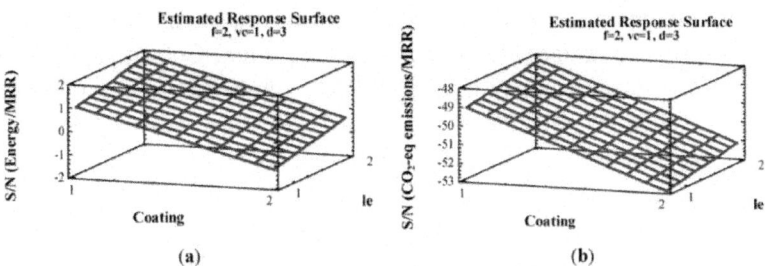

Figure 8:. (a) Values of the S/N ratio (Energy/MRR) with respect to *coating* and *le*; (b) Values of the S/N ratio (CO_2-eq emissions/MRR) with respect to *coating* and *le*.

Therefore, linear regression can explain the relationship between S/N Energy/MRR) and S/N (CO_2-eq emissions/MRR). The output shows the results of fitting a linear model via Equation (2). The correlation is 0.9781, and R^2 is 95.66%. Figure 9shows the model of the fitted linear regression. The calculation of the energy can be used to determine CO_2-eq emissions without considering the contribution of tools:MRR)]

$$S/N\,(CO_2 - eq\,emissions/MRR) = -53.74 + 0.93 \times [S/N(Energy/MRR)] \quad (2)$$

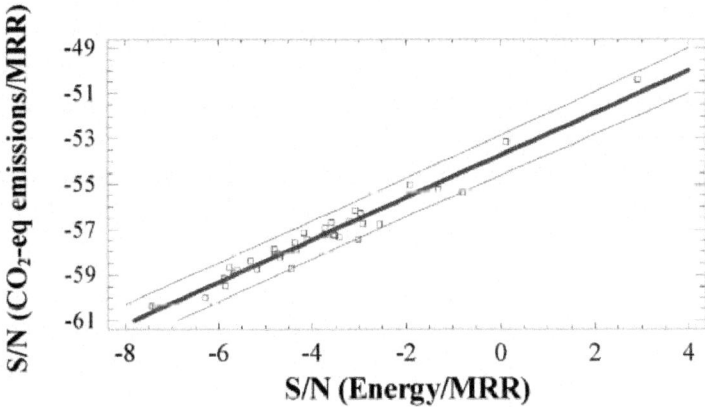

Figure 9:. Fitted linear regression mol.

CONCLUSIONS

We have presented a study centered on the selection of cutting inserts for environmentally efficient machining. This paper contributes to the understanding of this field by providing a quantitative evaluation of the main different technological and operational factors associated with modern machining using cutting inserts. Utilizing a widely used titanium alloy, the approach includes proper qualitative ratios of energy use and CO_2 impact related to a main manufacturing activity metric (MRR). Comprehensive experimentation and a subsequent statistical analysis using the Taguchi and ANOVA techniques represent a systematic methodology for qualitative and quantitative assessment. The methodology proposed can contribute to the conservation of resources through the identification of practices in manufacturing processes, particularly in the turning operations of Ti6Al4V alloy, that result in lower energy/MRR consumption and lower CO_2-eq emissions/MRR. For both metrics, energy and emissions, the study outcomes show that a high tool cutting length and a high cutting depth were identified as significant factors. Thus, this alloy is sensitive to cutting conditions and also to tool geometry, rather than tool material.

The inserts were ranked from those that have minor effects on the energy/MRR and CO_2-eq emissions/MRR to those that have major effects in these ratios: CNMG, CNMG PVD, WNMG PVD and WNMG. The coating has a low impact and an uncoated insert is a first choice for a high cutting depth and a high tool cutting length. This is a remarkable conclusion, because the use of coated inserts can be avoided in roughing operations, improving cost and performance.

This study demonstrates the need to consider cutting inserts as an element capable of reducing the machining contribution to climate change through the industrial activity. The carbon footprint is directly proportional to energy consumption. In this sense, forthcoming studies related to the sustainability and environmental impact of machining processes could focus more on the fundamental analysis of energy rather than the analysis of the downstream effect of emissions.

ACKNOWLEDGMENTS

The authors thank the "Research Group Industrial Production and Manufacturing Engineering (IPME)", the Spanish Ministry of Economy and Competitiveness through the DPI2011-27135 and DPI2014-58007-R projects and the Industrial Engineering School-UNED through the REF2014-ICF07 project for their superest.

REFERENCES

1.   Intergovernmental Panel on Climate Change (IPCC). *Climate Change 2014, Report*; IPCC: Geneva, Switzerland,holar]

2.   Rosen, M.A. Engineering and sustainability: Attitudes and actions. *Sustainability* 2013, 5, 37cssRef]

3.   Tanaka, K. Review of policies and measures for energy efficiency in industry sector. *Energy Policy* 2011, *39*, 6532cssRef]

4.   Zhai, Q.; Cao, H.; Zhao, X.; Yuan, C. Cost benefit analysis of using clean energy supplies to reduce greenhouse gas emissions of global automotive manufacturing. *Energies* 2011, *4*, 1478cssRef]

5.   Tan, X.C.; Wang, Y.Y.; Gu, B.H.; Mu, Z.K.; Yang, C. Improved methods for production manufacturing processes in environmentally benign manufacturing. *Energies* 2011, *4*, 1391cssRef]

6.   Domingo, R.; García, M.; Sánchez, A.; Gómez, R. A sustainable evaluation of drilling parameters for PEEK-GF30.*Materials* 2013, *6*, 5907cssRef]

7.   Beggs, P.J. Adaptation to impacts of climate change on aeroallergens and allergic respiratory diseases. *Int. J. Environ. Res. Public Health* 2010, *7*, 3006cssRef] [PubMed]

8.   Carlin, A. A Multidisciplinary, science-based approach to the economics of climate change. *Int. J. Environ. Res. Public Health* 2011, *8*, 985cssRef] [PubMed]

9.   Aguado, S.; Alvarez, R.; Domingo, R. Model of efficient and sustainable improvements in a lean production system through processes of environmental innovation. *J. Clean. Prod.* 2013, *47*, 14cssRef]

10. Nishitani, K.; Kaneko, S.; Fujii, H.; Komatsu, S. Are firms' voluntary environmental management activities beneficial for the environment and business? An empirical study focusing on Japanese manufacturing firms. *J. Environ. Manag.*2012, *105*, 12cssRef] [PubMed]

11. Yoon, H.S.; Kim, E.S.; Kim, M.S.; Lee, J.Y.; Lee, G.B.; Ahn, S.H. Towards greener machine tools—A review on energy saving strategies and technologies. *Renew. Sustain. Energy Rev.* 2015, *48*, 87cssRef]

12. Hu, S.; Liu, F.; He, Y.; Hu, T. An on-line approach for energy efficiency monitoring of machine tools. *J. Clean. Prod.*2012, *27*, 13cssRef]

13. Avram, O.I.; Xirouchakis, P. Evaluating the use phase energy requirements of a machine tool system. *J. Clean. Prod.*2011, *19*, 69cssRef]

14. Sreejith, P.S.; Ngoi, B.K.A. Dry machining: Machining of the future. *J. Mater. Process. Technol.* 2000, *101*, 28cssRef]

15. Pusavec, F.; Krajnik, P.; Kopac, J. Transitioning to sustainable production—Part I: Application on machining technologies. *J. Clean. Prod.* 2010, *18*, 17cssRef]

16. Pusavec, F.; Kramar, D.; Krajnik, P.; Kopac, J. Transitioning to sustainable production—Part II: Evaluation of sustainable machining technologies. *J. Clean. Prod.* 2010, *18*, 1211cssRef]

17. Kalpakjian, S.; Schmid, S.R. *Manufacturing Engineering and Technology*, 7th ed.; Prentice Hall: Upper Saddle River, NJ, USA,holar]

18. Alvarez, R.; Domingo, R.; Sebastian, M.A. The formation of saw toothed chip in a titanium alloy: Influence of constitutive models. *Strojniski Vestn. J. Mech. Eng.* 2011, *57*, 73cssRef]

19. Balazic, M.; Kopac, J. Machining of titanium alloy Ti-6Al-4V for biomedical applications. *StrojniskiVestn. J. Mech. Eng.*2010, *56*holar]

20. Norgate, T.E.; Wellwood, G. The potential applications for titanium metal powder and their life cycle impacts. *Jom*2006, *58*, cssRef]

21. Kundrak, J.; Mamalis, A.G.; Gyani, K.; Markopoulos, A. Environmentally friendly precision machining. *Mater. Manuf. Process.* 2006, *21*, cssRef]

22. Özel, T.; Sima, M.; Srivastava, A.K.; Kaftanoglu, B. Investigations on the effects of multi-layered coated inserts in machining Ti-6Al-4V alloy with experiments and finite element simulations. *CIRP Ann. Manuf. Technol.* 2010, *59*, cssRef]

23. Rajemi, M.F.; Mativenga, P.T.; Aramcharoen, A. Sustainable machining: Selection of optimum turning conditions based on minimum energy considerations. *J. Clean. Prod.* 2010, *18*, 1059cssRef]

24. Balogun, V.A.; Mativenga, P.T. Modelling of direct energy requirements in mechanical machining processes. *J. Clean. Prod.* 2013, *41*, 17cssRef]

25. Peng, T.; Xu, X.; Wang, L. A novel energy demand modelling approach for CNC machining based on function blocks.*J. Manuf. Syst.* 2014, *33*, 19cssRef]

26. Iqbal, A.; Zhang, H.C.; Kong, L.L.; Hussain, G. A rule-based system for trade-off among energy consumption, tool life, and productivity in machining process. *J. Intell. Manuf.*cssRef]

27. Schultheiss, F.; Zhou, J.; Gröntoft, E.; Stahl, J.E. Sustainable machining through increasing the cutting tool utilization.*J. Clean. Prod.* 2013, *59*, 29cssRef]

28. Mativenga, P.T.; Rajemi, M.F. Calculation of optimum cutting parameters based on minimum energy footprint. *CIRP Ann. Manuf. Technol.* 2011, *60*, 14cssRef]

29. Fang, K.; Uhan, N.; Zhao, F.; Sutherland, J.W. A new approach to scheduling in manufacturing for power consumption and carbon footprint reduction. *J. Manuf. Syst.* 2011, *30*, 23cssRef]

30. Yingjie, Z. Energy efficiency techniques in machining process: A review. *Int. J. Adv. Manuf. Technol.* 2014, *71*, 1123cssRef]

31. Peng, T.; Xu, X. Energy-efficient machining systems: A critical review. *Int. J. Adv. Manuf. Technol.* 2014, *72*, 1389cssRef]

32. Matweb. Material Propriety Data. Available online: http://www.matweb. com (accessed on 15 May 2013).

33. International Organization for Standardization (ISO). *ISO 1832:2012. Indexable Inserts for Cutting Tools—Designation, International Organization for Standardization*; ISO: Geneva, Switzerland,holar]

34. Klocke, F.; Döbbeler, B.; Binder, M.; Kramer, N.; Grüter, R.; Lung, D. Ecological evaluation of PVD and CVD coating systems in metal cutting processes assessed. In Proceedings of the 11th Global Conference on Sustainable Manufacturing—Innovative Solutions, Berlin, Germany, 23–25 September 2013; Seliger, G., Ed.; pp. 381–386.

35. Andriya, N.; Rao, P.V.; Ghosh, S. Dry machining of Ti-6Al-4V using PVD coated TiAlN tools. In Proceedings of the World Congress on Engineering, London, UK, 4–6 July 2012; pp. 1–6.

36. Prengel, H.G.; Jindal, P.C.; Wendt, K.H.; Santhanam, A.T.; Hegde, P.L.; Penich, R.M. A new class of high performance PVD coatings for carbide cutting tools. *Surf. Coat. Technol.* 2001, *139*, cssRef]

37. Jaffery, S.I.; Mativenga, P.T. Assessment of the machinability of Ti-6Al-4V alloy using the wear map approach. *Int. J. Adv. Manuf. Technol.* 2009, *40*, 68cssRef]

38. Dahmus, J.B.; Gutowski, T.G. An environmental analysis of machining. In Proceedings of the American Society of Mechanical Engineers (ASME) International Mechanical Engineering Congress, Anaheim, CA, USA, 13–19 November 2004; pp. 1–10.

39. Sun, S.; Brandt, M.; Mo, P.T.M. Evolution of tool wear and its effect on cutting forces during dry machining of Ti-6Al-4V alloy. *Proc. Inst. Mech. Eng. BJ. Eng. Manuf.* 2014, *228*, 19cssRef]

40. Red Eléctrica Española (RED). *Informe Mensual*; Red Eléctrica Española (RED): Alcobendas, Madrid, Spain,holar]

41. Intergovernmental Panel on Climate Change. Volume 3: Industrial processes and product use, Chapter 4: Metal industry emissions. In *IPCC Guidelines for National Greenhouse Gas Inventories*; Eggleston, H.S., Buendia, L., Miwa, K., Ngara, T., Tanabe, K., Eds.; Institute for Global Environmental Strategies: Kanagawa, Japan,holar]

42. Pré Consultants. *SimaPro 8 Software*; Pré Consultants: Amersfoort, The Netherlands,holar]

43. International Organization for Standardization (ISO). *ISO 3685:1993. Tool-Life Testing with Single-Point Turning Tools*; ISO: Geneva, Switzerland,holar]

44. Taguchi, G. *Introduction to Quality Engineering*; Asian Productivity Organization: Tokyo, Japan,holar]

45. Kopac, J.; Bahor, M.; Sokovic, M. Optimal machining parameters for achieving the desired surface roughness in fine turning of cold pre-formed steel workpieces. *Int. J. Mach. Tools Manuf.* 2002, *42*, 70cssRef]

46. Motorcu, A.R. The optimization of machining parameters using the Taguchi method for surface roughness of AISI 8660 hardened alloy steel. *StrojniskiVestn. J. Mech. Eng.* 2010, *56*, 39holar]

47. Montgomery, D.C. *Design and Analysis of Experiments*, 5th ed.; John Wiley & Sons Inc.: Hoboken, NJ, USA,holar]

48. Statgraphics. Available online: http://www.statgraphics.com/ (accessed on 22 September 2014).

49. Bhushan, R.K. Optimization of cutting parameters for minimizing power consumption and maximizing tool life during machining of Al alloy SiC particle composites. *J. Clean. Prod.* **2013**, *39*, 24csRef]

Chapter 5

MACHINING PERFORMANCE OF SPUTTER-DEPOSITED $(Al_{0.34}Cr_{0.22}Nb_{0.11}Si_{0.11}Ti_{0.22})_{50}N_{50}$ HIGH-ENTROPY NITRIDE COATINGS

Wan-Jui Shen[1], Ming-Hung Tsai[2], and Jien-Wei Yeh[1]

[1]Department of Materials Science and Engineering, National Tsing Hua University, Hsinchu 30013, Taiwan

[2]Department of Materials Science and Engineering, National Chung Hsing University, Taichung 40227, Taiwan

ABSTRACT

$(Al_{0.34}Cr_{0.22}Nb_{0.11}Si_{0.11}Ti_{0.22})_{50}N_{50}$ high-entropy nitride coatings prepared by reactive magnetron sputtering have been proved to have high hardness and superior oxidation resistance. Their thermal stability, adhesion strength, and cutting performance were investigated in this study. Hardness of the coating is 36 GPa, which only decreases slightly to 33 GPa after 900 °C annealing either in air or in vacuum for 2 h. No significant change in phase and microstructure were detected after annealing at 1000 °C. Rockwell C indentation and scratch tests shows that Ti interlayer provides a good adhesion between the nitride film and WC/Co substrates. In various milling tests, inserts coated with $(Al_{0.34}Cr_{0.22}Nb_{0.11}Si_{0.11}Ti_{0.22})_{50}N_{50}$ have evidently smaller flank wear depth than commercial inserts coated with TiN and TiAlN, even with their smaller thickness. Therefore, the $(Al_{0.34}Cr_{0.22}Nb_{0.11}Si_{0.11}Ti_{0.22})_{50}N_{50}$ coating has great potential in hard coating applications.

INTRODUCTION

Conventionally, high speed machining is usually performed with cutting fluids. Cutting fluids not only serve as lubricants but also dissipate massive heat generated in high speed machining. However, cutting fluids are not environmental friendly and meanwhile raise the cost of machining. Thus, dry machining are gradually being adopted in industry [1,2,3,4].

Some research indicates that carbide inserts survive longer in dry machining; still the heat generated during high speed machining will cause oxidation and coarsening and thus deteriorates the quality of inserts. The application of hard coatings, either by physical or chemical vapor deposition (PVD or CVD), is a common solution to prolong tool life [5,6].

Nitride coatings deposited using PVD techniques, such as TiN, TiCN, or TiAlN, are of special interest for their impressive performance and relatively lower costs. TiN is the most studied and used coating system for its high hardness, low coefficient of friction, and good adhesion to the substrates. However, its rapid oxidation starting at 550 °C limits the application in high speed machining [7,8]. TiCN coatings are known for their self-lubricating property and the particular low coefficient of friction [9,10,11,12]. However, their insufficient thermal stability and oxidation resistance make them inadequate for high temperature use. TiAlN is also a widely applied system. The addition of Al increases coating hardness significantly [8,13]. Moreover, the formation of dense Al_2O_3 layer at high temperature effectively prevents the inward diffusion of oxygen, and also enhances the hot hardness and chemical stability of coatings [14,15]. Higher cutting speed, better machining quality, and longer tool life are thus accomplished by the usage of TiAlN coatings [13,16]. However, severe oxidation of TiAlN still takes place when temperature is higher than 800 °C [7,17]. Extended systems based on these coatings are thus developed and studied successively to meet the demands for higher temperature applications.

A new category of nitride coatings based on high-entropy alloys (HEAs) [18], *i.e.*, high-entropy nitrides (HENs), has been studied in recent years. These HEN coatings, with five or more principal target elements, are reported to possess many attractive properties. For example, the hardness of $(AlCrTaTiZr)_{100-x}N_x$ [19,20], $(AlCrSiTiV)_{100-x}N_x$ [21], $(AlCrMoSiTi)_{100-x}N_x$ [22], and $(AlMoNbSiTaTiVZr)_{100-x}N_x$ [23], $(TiVCrZrHf)_{100-x}N_x$ [24], and $(TiZrNbHfTa)_{100-x}N_x$ [25] all fall in the range of 32–36 GPa when the value of x is about 50. The hardness of $(AlCrNbSiTiV)N$ is even higher (42 GPa), and still remains in the superhard level after vacuum annealing at 1000 °C for 5 h [26,27]. Some HENs not only have high hardness, but also exhibit excellent oxidation resistance. For example, non-equimolar $(Al_{0.34}Cr_{0.22}Nb_{0.11}Si_{0.11}Ti_{0.22})_{50}N_{50}$ coatings has a hardness of 36 GPa, and the thickness of surface oxide after air annealing at 900 °C for 50 h is only 290 nm [28,29]. Nice thermal stability of $(AlCrTaTiZr)N_x$ has been reported by Chang *et al.* [30] and Lai *et al.* [19] reported pleasing anti-wear performances demonstrated in $(AlCrTaTiZr)_{100-x}N_x$. These properties make HENs very promising in tool protection. However, no results of practical cutting tests are reported so far.

In this study, the thermal stability and machining properties of $(Al_{0.34}Cr_{0.22}Nb_{0.11}Si_{0.11}Ti_{0.22})_{50}N_{50}$ coatings are investigated in consideration of its excellent oxidation resistance and high hardness mentioned above. Firstly, X-ray diffraction (XRD) and nanoindentation are used to study the crystallographic structure and hardness of annealed coatings. Furthermore, various interlayers are applied to enhance coatings' adhesion to the WC/Co substrates. Lastly, the machining properties of the HEN coated inserts are tested by face milling of 304 stainless steel and SKD11 steel. The performance of commercial TiN and TiAlN coated inserts are also evaluated for comparison.

MATERIALS AND METHODS

Target and Film Preparation

The procedure of target preparation has been described in previous literature. The $(Al_{0.34}Cr_{0.22}Nb_{0.11}Si_{0.11}Ti_{0.22})_{50}N_{50}$ coatings were deposited on oxidized (100) silicon wafer which has a 200 nm-thick SiO_2 layer on surface, WC/Co substrates (2 cm × 2 cm), and cemented carbide inserts (Hitachi TPMN160308 EX35). The deposition was conducted in a direct current (DC) magnetron sputtering system with a background pressure below 1.33×10^{-3} Pa (1×10^{-5} Torr.) from $Al_{0.34}Cr_{0.22}Nb_{0.11}Si_{0.11}Ti_{0.22}$ alloy target. Reactive sputtering was performed at a fixed working pressure of 0.667 Pa (5 mTorr) in an Ar/N_2 gas mixture. The gas flow rate is 20 sccm for both Ar and N_2. Film was grown to a thickness about 1.4 μm. Interlayers were deposited onto WC/Co substrates from Cr, Ti, and $Al_{0.34}Cr_{0.22}Nb_{0.11}Si_{0.11}Ti_{0.22}$ targets before depositing nitride coatings. The interlayers were deposited in pure Ar atmosphere at a flow rate of 40 sccm at working pressure of 0.667 Pa. The thickness is controlled at about 100 nm. Substrate temperature and sputtering power was maintained at 415 °C and 150 W, respectively, throughout the deposition. The working distance was about 11 cm. Both of the alloy target and the substrates are pre-sputtered before the deposition to avoid contamination.

Film Characterization

The chemical composition of these (Al, Cr, Nb, Si, Ti)N coatings was analyzed using a field-emission electron probe micro-analyzer (FE-EPMA, JXA-8500F, JEOL, Tokyo, Japan). The coating is a stoichiometric nitride, and the composition of the target elements is close to the designed ratio. The formula of this HEN can thus be denoted as $(Al_{0.34}Cr_{0.22}Nb_{0.11}Si_{0.11}Ti_{0.22})_{50}N_{50}$. The vacuum annealing of samples was performed in a rapid thermal annealing furnace with a background pressure lower than 1.33×10^{-4} Pa (1×10^{-6} Torr.). The temperature was raised at ramping rate of 200 °C/min to targeting temperature.

After holding for 2 h, the sample was furnace-cooled to room temperature. Air annealing was conducted in an air furnace at various temperatures for 2 h with ramping rate of 15 °C/min. The crystallographic structures of coatings were characterized using a glancing angle X-ray diffractometer (GIXRD, MXP18, MAC Science, Japan) with Cu Kα radiation operated at 40 kV and 150 mA. The scanning speed was 4 °/min and the incident angle was 1°. The hardness values of films deposited on (100) Si wafers were measured with a Micromaterials Nano Test indentation system using a Berkovich indenter and a continuous applied load of 5 mN. The roughness of the HEN films examined by AFM (atomic force microscope, NS3a-controller with D3100 stage, Digital Instrument, USA) is lower than 0.5 nm. The indentation depth was controlled to be less than 1/10 of the film thickness to avoid substrate effect. The hardness was calculated from the loading/uploading curves of indentation tests following the analysis method proposed by Oliver and Pharr [31]. Ten points were measured for each sample to obtain a reliable result. A field-emission scanning electron microscope (FESEM, JSM-6500F, Tokyo, Japan) operated at 15 kV was used to observe surface and cross-sectional microstructures of films.

Adhesion Test

The adhesion of films was examined by both Rockwell C indentation and scratch test. In Rockwell C indentation test, a 120° cone-shaped diamond tip was applied perpendicularly to the film surface with a load of 1470 N (150 kgf). The morphology of indentation area was observed using an optical microscope (OM, Pentad Scientific Corporation, Hsinchu, Taiwan), and then classified into six ratings evaluating the damage degree from HF1 to HF6 [32]. In this classification, HF1 indicates the least cracks near the indentation boundaries and implies a good bonding of films to the substrates. HF6 represents severe delamination of the coatings, and indicates a poor adhesion of films. Adhesions rated from HF1 to HF4 are adequate in commercial demands. The scratch test was performed using a progressive load scratch tester (Sense Ted, Ltd., Kaohsiung, Taiwan) equipped with a spherical Rockwell diamond indenter (200 μm in radius). The indenter was slid over coatings surface and the load was increased from 0 to 100 N. The loading rate was 100 N/min, while the sliding speed is 10 mm/min. Five scratches were made for each sample to get a trusted result. The critical load corresponding to adhesive failure on the scratch track was evaluated by OM observation. Stallard *et al.* [33] identified four kinds of film failure events and labeled as critical loads from L_{C1} to L_{C4}. L_{C1} and L_{C2} belong to the cohesive failure mode, which is more related to the coating's intrinsic properties. L_{C3} denotes the load at which the initial adhesive failure occurs, while L_{C4} indicates the total failure of coating that the substrate has been exposed to completely. Thus, L_{C3} and L_{C4} provide a better index for evaluating the adhesion quality of coatings. In practical use, the critical loads L_{C3} and L_{C4} should be higher than 60 N to meet the industrial demands.

Milling Test

The machining properties of TiN, TiAlN, and HEN coated inserts were evaluated by dry face milling of SKD11 steel and 304 stainless steel. The hardness of SKD11 steel is 260 Hv, while that of 304 Stainless steel is 190 Hv. Tests on 304 stainless steel were performed at two cutting speeds: 76 and 160 m/min. Feed rate was 0.22 mm/rev. Depth of cut was controlled at 0.2 mm for the lower cutting speed and 0.1 mm for the higher one to avoid severe vibration of the milling machine. For SKD11 steel, the cutting speed was 76 mm/min. Depth of cut was 0.1 mm, and the feed rate was 0.22 mm/rev. Each milling was interrupted every 180 m of cutting distance and the morphology of the rake face was observed using SEM so that the flank wear can be measured. For comparison, a continuous milling test of 304 stainless steel at higher cutting speed, 160 m/min, was also undertaken to imitate the practical use of inserts and evaluate the continuous protection ability of different coatings at higher cutting temperature.

RESULTS AND DISCUSSION

Thermal Stability and Film Hardness

The high hardness and outstanding oxidation resistance of $(Al_{0.34}Cr_{0.22}Nb_{0.11}Si_{0.11}Ti_{0.22})_{50}N_{50}$ have already been reported previously [29]. In practical machining operations, however, the coatings will be heated to temperatures as high as 1000 °C during machining, and then cooled to room temperature. Therefore, the stability of the structure and properties of the coatings after these excessive heating is critical and shall be tested. Figure 1 shows the XRD patterns of the coating after vacuum annealing at different temperatures. All coatings retain their NaCl-type FCC structure. Additionally, no significant grain growth is observed—even for samples annealed at 1000 °C. The suppression of grain coarsening in high-entropy nitride at high temperature has been reported by Huang et al. [26]. The reason for the stability of phase and structure at high temperature is twofold. Firstly, the stability of solid solution phase at high temperature is the major reflection of high-entropy effect. Secondly, severe lattice distortion effect reduces grain boundary energy and thus lowers the driving force of grain coarsening [26]. It is also noted that a small shift of peaks indicating the decrease of lattice constant is found as annealing temperature increases. This reduction is attributed to the elimination of point defects introduced during sputtering deposition.

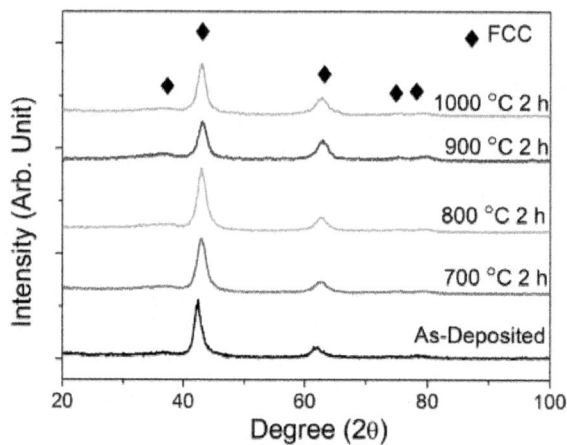

Figure 1: XRD patterns of as-deposited and vacuum-annealed $(Al_{0.34}Cr_{0.22}Nb_{0.11}Si_{0.11}Ti_{0.22})_{50}N_{50}$ coatings.

The hardness of coatings after air or vacuum annealing is shown in Figure 2. As-deposited coatings have a hardness of 36 GPa. The high hardness of the as-deposited $(Al_{0.34}Cr_{0.22}Nb_{0.11}Si_{0.11}Ti_{0.22})_{50}N_{50}$ coatings was reported to be a combined result of nano-sized grains, dense structures, and residual stress [28]. Annealing at 700, 800, and 900 °C in vacuum slightly softens the coatings, but the hardness values still remain higher than 33 GPa. As long as the change in structure due to annealing is insignificant (700–900 °C), the strengthening effect is still retained. In addition, the excellent capability of high-entropy nitrides to retain the original structure at high temperatures is also reported by Huang, Lai, and Chang [26,34,35]. Annealing at 1000 °C decreases the film hardness value to 26 GPa, but this value is still higher than that of commercial TiN coatings [36]. This decrease of hardness might be due to the loss of nitrogen in vacuum. On the other hand, all coatings annealed in air have hardness values between 31 and 33 GPa, which means no significant change in hardness takes place. It was reported that a thin, multi-layer oxide scale forms on the surface of the $(Al_{0.34}Cr_{0.22}Nb_{0.11}Si_{0.11}Ti_{0.22})_{50}N_{50}$ coatings after annealing. This dense oxide scale acts as a diffusion barrier so that the inner nitride is not further oxidized [29]. The outward diffusion of nitrogen is also inhibited due to the oxide scale. Therefore, the hardness is preserved.

The above demonstrates that both the structure and hardness of the $(Al_{0.34}$ $Cr_{0.22}Nb_{0.11}Si_{0.11}Ti_{0.22})_{50}N_{50}$ coatings are not greatly affected after heat treatment. Such outstanding stability guarantees the quality of the coatings in practical machining operations.

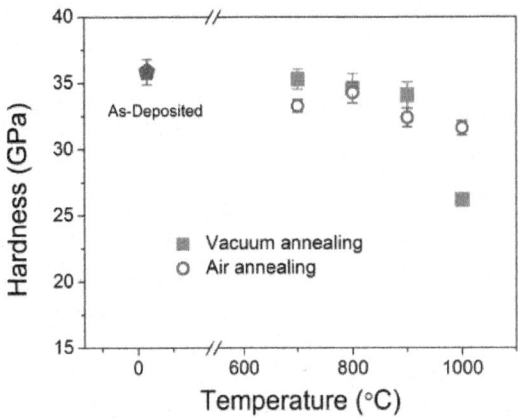

Figure 2: Hardness values of as-deposited and air/vacuum-annealed $(Al_{0.34}Cr_{0.22}Nb_{0.11}$ $Si_{0.11}Ti_{0.22})_{50}N_{50}$ coatings.

Adhesion Test

The adhesion of the coating to WC/Co substrates was studied by two most common adhesion tests, Rockwell C indentation test and progressive load scratch test. To optimize the adhesion of $(Al_{0.34}Cr_{0.22}Nb_{0.11}Si_{0.11}Ti_{0.22})_{50}N_{50}$ on WC/Co substrates, three different interlayers—Cr, Ti, and $Al_{0.34}Cr_{0.22}Nb_{0.11}Si_{0.11}Ti_{0.22}$— were applied on the substrates before the deposition of HEN coating.

The evaluation of adhesion strength by Rockwell C indentation tests is based on the damages of the coatings adjacent to the indentation boundary, and six levels of adhesion, HF1 to HF6, have been proposed previously [32]. HF1 to HF4 indicate adequate adhesion, and no evident spallation is seen around the boundary. In contrast, for HF5 and HF6, severe delamination has taken place, which reflects inadequate adhesion between film and substrate. Based on the images of films after Rockwell C tests (Figure 3), Cr interlayer is incapable of providing some adhesion (HF5), while Ti and $Al_{0.34}Cr_{0.22}Nb_{0.11}Si_{0.11}Ti_{0.22}$ interlayers provide better adhesion strength (HF4). Also, the degree of cracking represents the toughness of coatings to some extent. Cracks found at the boundary of indentations illustrate a medium toughness of $(Al_{0.34}Cr_{0.22}Nb_{0.11}Si_{0.11}Ti_{0.22})_{50}N_{50}$ coating itself.

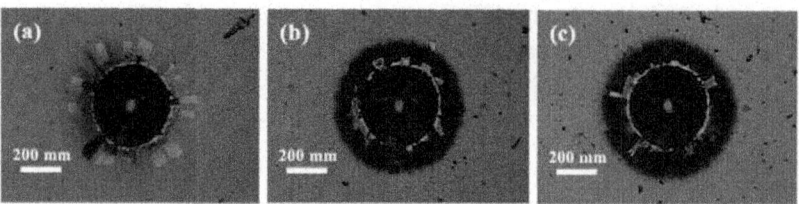

Figure 3: The OM images of Rockwell C indentation of $(Al_{0.34}Cr_{0.22}Nb_{0.11}Si_{0.11}Ti_{0.22})_{50}N_{50}$ coatings on WC/Co substrates with various interlayer: (a) Cr interlayer, (b) $Al_{0.34}Cr_{0.22}Nb_{0.11}Si_{0.11}Ti_{0.22}$ interlayer, and (c) Ti interlayer.

The morphology of tracks after scratch tests is presented in Figure 4. It is seen that the adhesion strength of Cr interlayer is poor. Its L_{C3} is only 3 N, and the L_{C4} is only 14 N. These values are apparently lower than that of the commercial criterion (≥ 60 N). HEN coatings using $Al_{0.34}Cr_{0.22}Nb_{0.11}Si_{0.11}Ti_{0.22}$ as interlayer exhibit higher critical load. The L_{C3} and L_{C4} are 47 and 55 N, respectively, but the values are still lower than 60 N. Ti interlayer provides the best adhesion among all interlayers, with both L_{C3} and L_{C4} values higher than 100 N (no delamination/cracks along the track is observed). The adhesion strengths based on scratch tests are consistent with that based on Rockwell-C indentation tests. These results are summarized in Table 1.

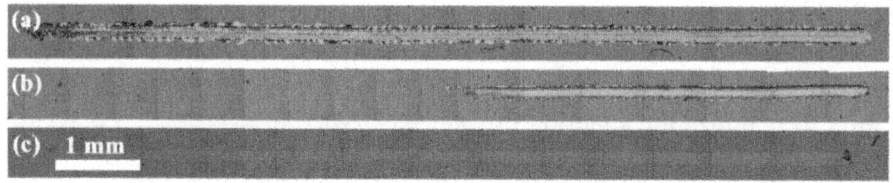

Figure 4: The OM images of the scratch track of $(Al_{0.34}Cr_{0.22}Nb_{0.11}Si_{0.11}Ti_{0.22})_{50}N_{50}$ coatings on WC/Co substrates with various interlayer: (a) Cr interlayer, (b) $Al_{0.34}Cr_{0.22}Nb_{0.11}Si_{0.11}Ti_{0.22}$ interlayer, and (c) Ti interlayer.

Table 1: Critical loads in scratch test and damage ratings in Rockwell C indentation of $(Al_{0.34}Cr_{0.22}Nb_{0.11}Si_{0.11}Ti_{0.22})_{50}N_{50}$ coatings with various interlayers

Interlayer	L_{C3}	L_{C4}	Rockwell C (150 kg)
Cr	2 ± 0.2 N	14 ± 4 N	HF5
$Al_{0.34}Cr_{0.22}Nb_{0.11}Si_{0.11}Ti_{0.22}$	47 ± 4 N	55 ± 2.9 N	HF4
Ti	>100 N	>100 N	HF4

Milling Test

In order to study the cutting performance of $(Al_{0.34}Cr_{0.22}Nb_{0.11}Si_{0.11}Ti_{0.22})_{50}N_{50}$ coatings, dry machining of 304 stainless steel and SKD11 steel is performed. The HEN coatings are deposited onto commercial cemented carbide inserts with 100 nm thick Ti interlayer. The cross-sectional and surface structures of the coatings on inserts are demonstrated in Figure 5a,b, respectively. The thickness of HEN coatings was measured from the SEM image and is about 1.4 μm. Its cross-sectional image shows a dense and fine fiber structure, while the surface consists of clusters of grains. For comparison, commercial TiN and TiAlN coated cemented carbide inserts are also tested. The measured hardness is 20 and 30 GPa for TiN and TiAlN coatings, respectively. The surface morphologies of these films are presented in Figure 5c,d, which reveal the typical surface structure of coatings deposited by the anodic vacuum arc process. XRD patterns of uncoated and coated inserts are shown inFigure 6. The uncoated insert exhibits peaks belonging to WC and Co. All three coatings have a simple B1 structure.

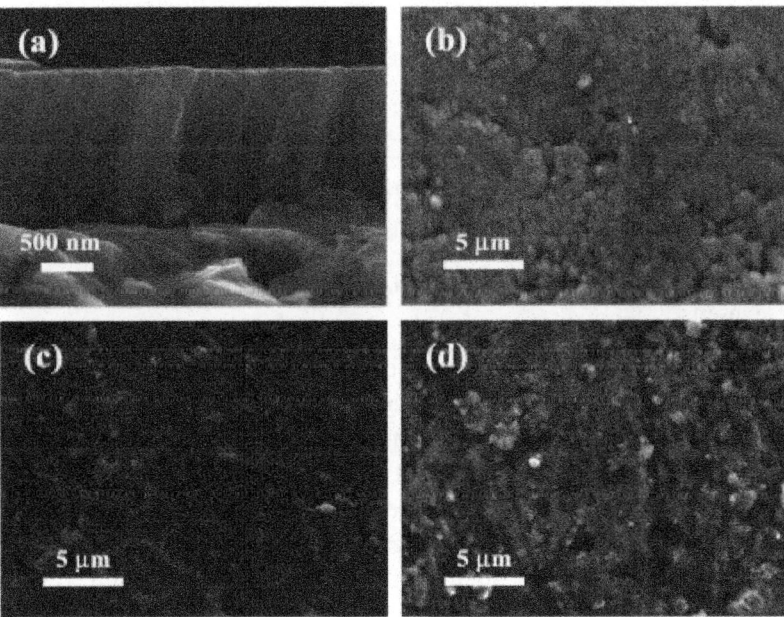

Figure 5: SEM micrographs of (a) cross-sectional and (b) surface structures of $(Al_{0.34}Cr_{0.22}Nb_{0.11}Si_{0.11}Ti_{0.22})_{50}N_{50}$ coatings; (c) surface structures of TiN coatings; and (d) surface structures of TiAlN coatings.

Insert coated with $(Al_{0.34}Cr_{0.22}Nb_{0.11}Si_{0.11}Ti_{0.22})_{50}N_{50}$ shows peaks of WC, indicating that the HEN film is thinner than TiN and TiAlN films. It is also noted that the relative intensities of peaks near 37° and 42° of HEN coatings are different in Figure 1 and Figure 6. This suggests that the orientation of film is influenced by substrate material since the substrate used for the films in Figure 1 was (100) Si with amorphous native-oxide surface layer SiO_2 whereas that used for the films in Figure 6 is crystalline WC/Co composite. Although the morphology and orientation of nuclei on the substrate surface could be related with the surface roughness, crystallinity and chemical composition, further investigation is needed in the future in order to provide a detailed explanation.

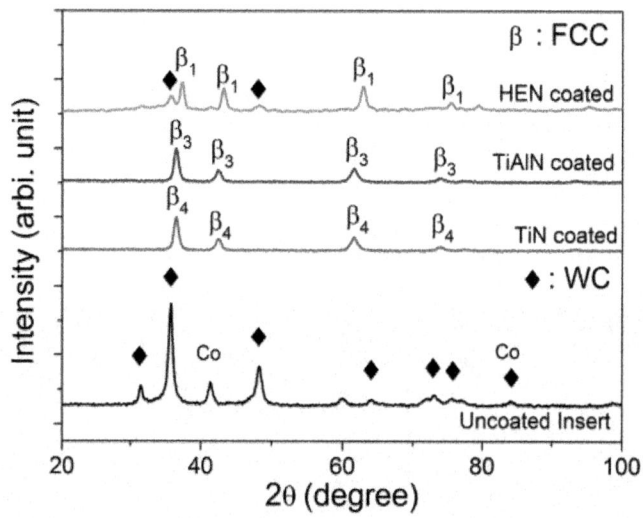

Figure 6: XRD patterns of uncoated, TiN, TiAlN, and $(Al_{0.34}Cr_{0.22}Nb_{0.11}Si_{0.11}Ti_{0.22})_{50}N_{50}$ coated inserts.

In machining tests, flank wear depth is a common index for tool life assessment. Lower flank wear depth indicates better wear resistance, and thus represents longer tool life. The maximum flank wear of insert edge is measured every 180 m of cutting distance, and the results for milling SKD11 steel are shown in Figure 7. In the initial stage (first 180 m of cutting distance) of the interrupted milling test, the HEN coating already demonstrates its advantage over TiN and TiAlN by its smallest flank wear depth of 90 μm. It should be mentioned that the thickness of HEN is smaller than that of TiN and TiAlN, which means HEN will further expand the lead over commercial counterparts if all coatings have the same thickness.

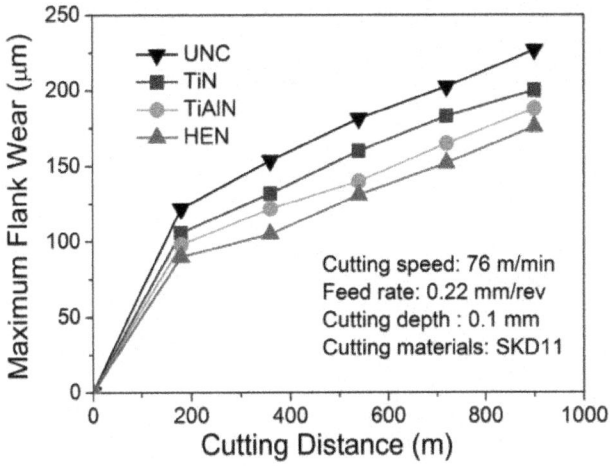

Figure 7: Maximum flank wear depths of uncoated, TiN, TiAlN, and $(Al_{0.34}Cr_{0.22}Nb_{0.11}Si_{0.11}Ti_{0.22})_{50}N_{50}$ coated inserts milling against SKD11 steel as a function of cutting distance in the interrupted milling test.

The HEN coating keeps its lead throughout the milling test. After a total cutting distance of 900 m, the wear depth of uncoated, TiN-, TiAlN-, and HEN-coated inserts are 227, 200, 188, and 176 μm, respectively. As shown in Figure 8, the wear lands of all inserts have similar appearance, and a small amount of debris is also found. The white region represents the exposed WC/Co substrate. Gray region represents the gradient wear of the film. The dark region is the un-worn film.

Figure 8: Flank wear patterns of cutting inserts coated with (a) TiN, (b) TiAlN, and (c) $(Al_{0.34}Cr_{0.22}Nb_{0.11}Si_{0.11}Ti_{0.22})_{50}N_{50}$ after interrupted milling tests against SKD11 steel with the cutting distance of 900 m.

Figure 9 demonstrates the interrupted milling results against 304 stainless steel (HV 190). At the smallest cutting distance (180 m), the three coated inserts have similar flank wears. However, for longer cutting distances, HEN-coated inserts has the smallest wear depth. After 900 m of cutting, the wear depth of uncoated, TiN-, TiAlN-, and HEN-coated inserts are 226, 202, 184, and 175 µm, respectively. HEN coatings again show advantage over the two commercial counterparts.

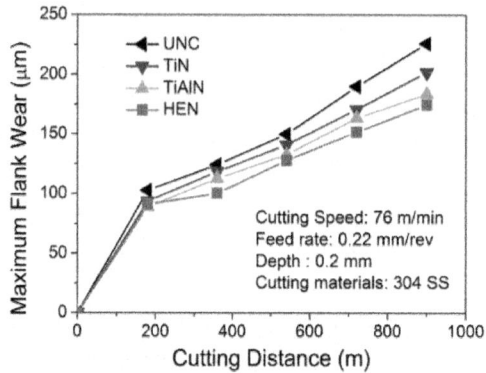

Figure 9: Maximum flank wear depths of uncoated, TiN, TiAlN, and $(Al_{0.34}Cr_{0.22}Nb_{0.11}Si_{0.11}Ti_{0.22})_{50}N_{50}$ coated inserts milling against 304 stainless steel as a function of cutting distance in the interrupted milling test.

The morphology of the wear land against 304 stainless steel is shown in Figure 10. The feature is quite different from that seen in Figure 10. This is because 304 stainless steel is tougher than SKD 11 workpiece used for Figure 8, and cutting depth is double that used for Figure 8. That means the present cutting condition is severer than that for Figure 8. The cutting edge has some worn-out regions and the wear land shows the exposed WC/Co substrate (grain-like and white). Some stainless steel layers (thick layer is gray and thin layer is white) adhere to the worn-out edge, exposed WC/Co substrate and even un-worn film. It is noted that HEN-coated insert also displays a smaller worn-out cutting edge than the other two indicating better protection.

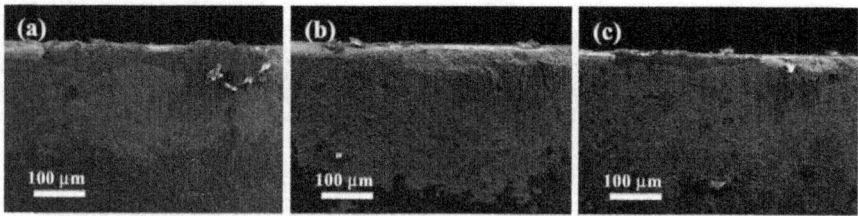

Figure 10: Flank wear patterns of cutting inserts coated with (a) TiN, (b) TiAlN, and (c) $(Al_{0.34}Cr_{0.22}Nb_{0.11}Si_{0.11}Ti_{0.22})_{50}N_{50}$ after interrupted milling tests against 304 stainless steel with the cutting distance of 900 m.

A more rigorous machining condition, *i.e.*, continuous machining at higher cutting speed (160 m/min), is executed. The inserts were observed using SEM until the cutting distance had reached 900 m. The difference in wear resistance between HEN-coated inserts and commercial inserts at this condition is evidently larger—The wear depth of HEN-coated inserts is 23% and 25% smaller than that of TiN-coated and TiAlN-coated inserts, respectively, as shown in Figure 11. It is emphasized again here that the thickness of the HEN coating is the smallest among the three. If the thicknesses of the three coatings are equal, HEN should further outperform the commercial counterparts.

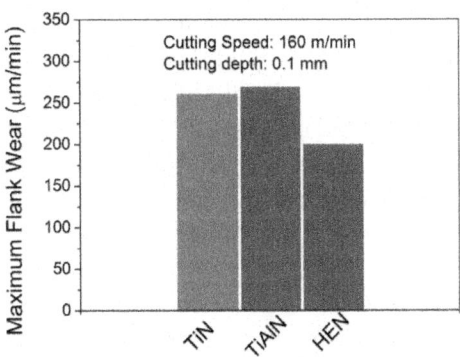

Figure 11: The comparison of maximum flank wear among TiN-, TiAlN-, and HEN-coated inserts after continuous milling tests against 304 stainless steel for a cutting distance of 900 m.

In such rigorous cutting conditions, the local temperature at the workpiece-tool contact point can be very high. This can significantly deteriorate the performance of the coatings due to two reasons. Firstly, the coating itself can soften and therefore lose its wear resistance. This is not only due to the intrinsic thermal softening of the material, but can also be a result of structural coarsening at high temperatures. Secondly, oxidation takes place at high temperature, which turns the nitride coatings to oxides with worse properties. Therefore, high structural stability and good oxidation resistance are critical to high speed cutting. We have already shown in Section 2.1 that the present HEN has remarkable stability. Annealing the HEN coating at 1000 °C for 2 h in air does not soften the coating significantly (36 GPa to 31 GPa). Note that this annealing temperature is very high and the annealing period is longer than the typical lifetime of tools. This suggests that under practical cutting conditions, the degree of softening should be smaller. For comparison, the hot hardness of TiN coating is reported to decrease from about 22.5 GPa at room temperature to 5.9 GPa at 1000 °C, and that of TiAlN decreases from about 25.0 to 11.8 GPa [13]. As for oxidation resistance, our previous research has already shown that the $(Al_{0.34}Cr_{0.22}Nb_{0.11}Si_{0.11}Ti_{0.22})_{50}N_{50}$ HEN possesses excellent quality in this regard. Note that annealing this HEN at 900 °C for 50 h only leads to an oxide layer 330 nm in thickness [29]. In contrast, TiN severely oxidizes at above 550 °C, while TiAlN starts rapid oxidation at above 800 °C [7,15]. The two advantages together lead to better cutting performances, particularly under harsh cutting conditions.

CONCLUSION

The thermal stability, adhesion strength and milling performance of the $(Al_{0.34}Cr_{0.22}Nb_{0.11}Si_{0.11}Ti_{0.22})_{50}N_{50}$ high-entropy nitride coating were studied. The HEN possesses good thermal stability. No significant change is seen in XRD patterns. The hardness also remains high—31 GPa after annealing at 900 °C for 2 h in air. Scratch tests show that among Ti, Cr, and $Al_{0.34}Cr_{0.22}Nb_{0.11}Si_{0.11}Ti_{0.22}$, Ti interlayer leads to the best adhesion strength, with both L_{C3} and L_{C4} values higher than 100 N. Inserts with HEN coatings and Ti interlayer show remarkable milling performances as compared to commercial TiN- and TiAlN-coated inserts. In milling tests both against SKD11 steel and 304 stainless steel, HEN-coated inserts have lower flank wear, even when HEN coating is thinner than TiN and TiAlN coatings. The cutting performance of HEN coating is even better as milling is operated continuously at higher speed. The advantage of the $(Al_{0.34}Cr_{0.22}Nb_{0.11}Si_{0.11}Ti_{0.22})_{50}N_{50}$ coating over TiN and TiAlN is a combined result of high hardness, good thermal stability and outstanding oxidation resistance. The $(Al_{0.34}Cr_{0.22}Nb_{0.11}Si_{0.11}Ti_{0.22})_{50}N_{50}$ coating with Ti interlayer hence shows great potential in protective coating applications.

ACKNOWLEDGEMENT

The authors gratefully acknowledge the financial support for this research from the National Science Council of Taiwan under grants NSC 100-2221-E-007-049-MY2.

AUTHOR CONTRIBUTIONS

Experimental measurements, analysis and interpretation of the results as well as conclusions were conducted by Wan-Jui Shen. The manuscript and artworks were prepared by Wan-Jui Shen and Ming-Hung Tsai, the whole paper was revised by Jien-Wei Yeh with approval by the other co-authors.

REFERENCES

1.   Derflinger, V.; Brandle, H.; Zimmermann, H. New hard/lubricant coating for dry machining. *Surf. Coat. Technol.* 1999, *113*, 286–292.

2.   Veprek, S.; Veprek-Heijman, M.J.G. Industrial applications of superhard nanocomposite coatings. *Surf. Coat. Technol.* 2008, *202*, 5063–5073.

3. Mayrhofer, P.H.; Mitterer, C.; Hultman, L.; Clemens, H. Microstructural design of hard coatings. *Prog. Mater. Sci.*2006, *51*, 1032–1114.

4. Kathrein, M.; Michotte, C.; Penoy, M.; Polcik, P.; Mitterer, C. Multifunctional multi-component PVD coatings for cutting tools. *Surf. Coat. Technol.* 2005, *200*, 1867–1871.

5. Knotek, O.; Löffler, F.; Krämer, G. Applications to cutting tools. In *Handbook of Hard Coatings*; Bunshah, R.F., Ed.; Noyes Publications: New York, NY, USA, 2001; pp. 370–410. [Google Scholar]

6. Knotek, O.; Lugscheider, E.; Loffler, F.; Kramer, G.; Zimmermann, H. Abrasive wear-resistance and cutting performance of complex PVD coatings. *Surf. Coat. Technol.* 1994, *68*, 489–493.

7. Wittmer, M.; Noser, J.; Melchior, H. Oxidation-kinetics of tin thin-films. *J. Appl. Phys.* 1981, *52*, 6659–6664.

8. Zhou, M.; Makino, Y.; Nose, M.; Nogi, K. Phase transition and properties of Ti-Al-N thin films prepared by RF-plasma assisted magnetron sputtering. *Thin Solid Films* 1999, *339*, 203–208.

9. Vancoille, E.; Celis, J.P.; Roos, J.R. Dry sliding wear of tin based ternary PVD coatings. *Wear* 1993, *165*, 41–49.

10. Polcar, T.; Novak, R.; Siroky, P. The tribological characteristics of TiCN coating at elevated temperatures. *Wear* 2006,*260*, 40–49.

11. Knotek, O.; Loffler, F.; Kramer, G. Acr deposition of Ti-C and Ti-C-N using acetylene as a reactive gas. *Vacuum* 1992,*43*, 645–648.

12. Erturk, E.; Knotek, O.; Burgmer, W.; Prengel, H.G.; Heuvel, H.J.; Dederichs, H.G.; Stossel, C. Ti(C,N) coatings using the arc process. *Surf. Coat. Technol.* 1991, *46*, 39–46.

13. PalDey, S.; Deevi, S.C. Single layer and multilayer wear resistant coatings of (Ti,Al)N: A review. *Mater. Sci. Eng. A*2003, *342*, 58–79.

14. Kawate, M.; Hashimoto, A.K.; Suzuki, T. Oxidation resistance of $Cr_{1-x}Al_xN$ and $Ti_{1-x}Al_xN$ films. *Surf. Coat. Technol.*2003, *165*, 163–167.

15. McIntyre, D.; Greene, J.E.; Hakansson, G.; Sundgren, J.E.; Münz, W.D. Oxidation of metastable single-phase polycrystalline $Ti_{0.5}Al_{0.5}N$ films: Kinetics and mechanisms. *J. Appl. Phys.* 1990, *67*, 1542–1553.

16. Jindal, P.C.; Santhanam, A.T.; Schleinkofer, U.; Shuster, A.F. Performance of PVD TiN, TiCN, and TiAlN coated cemented carbide tools in turning. *Int. J. Refract. Met. Hard Mater* 1999, *17*, 163–170.

17. Munz, W.D. Titanium aluminum nitride films—A new alternative to TiN coatings. *J. Vac. Sci. Technol. A* 1986, *4*, 2717–2725.

18. Tsai, M.H.; Yeh, J.W. High-entropy alloys: A critical review. *Mater. Res. Lett.* 2014, *2*, 107–123.

19. Lai, C.H.; Cheng, K.H.; Lin, S.J.; Yeh, J.W. Mechanical and tribological properties of multi-element (AlCrTaTiZr)N coatings. *Surf. Coat. Technol.* 2008, *202*, 3732–3738.

20. Lai, C.H.; Lin, S.J.; Yeh, J.W.; Davison, A. Effect of substrate bias on the structure and properties of multi-element (AlCrTaTiZr)N coatings. *J. Phys. D Appl. Phys.* 2006, *39*, 4628–4633.

21. Lin, C.H.; Duh, J.G.; Yeh, J.W. Multi-component nitride coatings derived from Ti-Al-Cr-Si-V target in rf magnetron sputter. *Surf. Coat. Technol.* 2007, *201*, 6304–6308.

22. Chang, H.W.; Huang, P.K.; Yeh, J.W.; Davison, A.; Tsau, C.H.; Yang, C.C. Influence of substrate bias, deposition temperature and post-deposition annealing on the structure and properties of multi-principal-component (AlCrMoSiTi)N coatings. *Surf. Coat. Technol.* 2008, *202*, 3360–3366.

23. Tsai, M.H.; Lai, C.H.; Yeh, J.W.; Gan, J.Y. Effects of nitrogen flow ratio on the structure and properties of reactively sputtered (AlMoNbSiTaTiVZr) N_x coatings. *J. Phys. D Appl. Phys.* 2008, *41*.

24. Tsai, D.C.; Chang, Z.C.; Kuo, L.Y.; Lin, T.J.; Lin, T.N.; Shieu, F.S. Solid solution coating of (TiVCrZrHf)N with unusual structural evolution. *Surf. Coat. Technol.* 2013, *217*, 84–87.

25. Braic, V.; Vladescu, A.; Balaceanu, M.; Luculescu, C.R.; Braic, M. Nanostructured multi-element (TiZrNbHfTa)N and (TiZrNbHfTa)C hard coatings. *Surf. Coat. Technol.* 2012, *211*, 117–121.

26. Huang, P.K.; Yeh, J.W. Inhibition of grain coarsening up to 1000 °C in (AlCrNbSiTiV)N superhard coatings. *Scr. Mater.* 2010, *62*, 105–108.

27. Huang, P.K.; Yeh, J.W. Effects of substrate bias on structure and mechanical properties of (AlCrNbSiTiV)N coatings. *J. Phys. D Appl. Phys.* 2009, *42*.

28. Shen, W.J.; Tsai, M.H.; Chang, Y.S.; Yeh, J.W. Effects of substrate bias on the structure and mechanical properties of $(Al_{1.5}CrNb_{0.5}Si_{0.5}Ti)N_x$ coatings. *Thin Solid Films* 2012, *520*, 6183–6188.

29. Shen, W.J.; Tsai, M.H.; Tsai, K.Y.; Juan, C.C.; Tsai, C.W.; Yeh, J.W.; Chang, Y.S. Superior oxidation resistance of $(Al_{0.34}Cr_{0.22}Nb_{0.11}Si_{0.11}Ti_{0.22})_{50}N_{50}$ high-entropy nitride. *J. Electrochem. Soc.* 2013, *160*, C531–C535.

30. Chang, S.Y.; Chen, M.K.; Chen, D.S. Multiprincipal-element AlCrTaTiZr-nitride nanocomposite film of extremely high thermal stability as diffusion barrier for Cu metallization. *J. Electrochem. Soc.* 2009, *156*, G37–G42.

31. Oliver, W.C.; Pharr, G.M. An improved technique for determing hardness and elastic-modulus using load and displacement sensing indentation experiments. *J. Mater. Res.* 1992, *7*, 1564–1583.

32. Cohen, M.L. Calculation of bulk moduli of diamond and zincblende solid. *Phys. Rev. B* 1985, *32*, 7988–7991.

33. Stallard, J.; Poulat, S.; Teer, D.G. The study of the adhesion of a TiN coating on steel and titanium alloy substrates using a multi-mode scratch tester. *Tribol. Int.* 2006, *39*, 159–166.

34. Lai, S.W. A study on nitride films of AlBCrSiTi high-entropy alloy by reactive dc sputtering. Master Thesis, National Tsing Hua University, Hsinchu, Taiwan, 2006. [Google Scholar]

35. Chang, S.Y.; Chen, D.S. 10-nm-thick quinary (AlCrTaTiZr)N film as effective diffusion barrier for Cu interconnects at 900 °C. *Appl. Phys. Lett.* 2009, *94*.

36. Sundgren, J.E. Structure and properties of tin coatings. *Thin Solid Films* 1985, *128*, 21–44.

Chapter 6

LASER MICROMACHINING OF GLASS, SILICON, AND CERAMICS

L. Rihakova[1] and H. Chmelickova[2]

[1]Palacky University, RCPTM, Joint Laboratory of Optics of Palacky University and Institute of Physics of the Academy of Sciences of the Czech Republic, 17. Listopadu 50a, 77207 Olomouc, Czech Republic

[2]Institute of Physics of the Academy of Sciences of the Czech Republic, Joint Laboratory of Optics of Palacky University and Institute of Physics of the Academy of Sciences of the Czech Republic, 17. Listopadu 50a, 77207 Olomouc, Czech Republic

ABSTRACT

A brief review is focused on laser micromachining of materials. Micromachining of materials is highly widespread method used in many industries, including semiconductors, electronic, medical, and automotive industries, communication, and aerospace. This method is a promising tool for material processing with micron and submicron resolution. In this paper micromachining of glass, silicon, and ceramics is considered. Interaction of these materials with laser radiation and recent research held on laser material treatment is provided.

INTRODUCTION

Miniaturization is an important trend in many modern technologies. Micromachining of materials with micron resolution at high speed is a widespread technology used in nearly all industries. This method can be found in manufacturing high-tech microproducts for biotechnological, microelectronics, telecommunication, MEMS, and medical applications. After absorption of radiation the material is removed by the process of laser ablation. Laser ablation usually relies on strong absorption of laser photons; thus the laser wavelength has to be appropriate unlike ultrafast lasers which are used. Ultrafast lasers cause ablation as a result of multiphoton absorption at high peak intensities, so that even materials transparent to the laser wavelength can be machined [1].

Many types of lasers are supposed to be used for micromachining materials. These include microsecond carbon dioxide lasers at wavelengths between 9.3 μm and 11 μm, nanosecond and femtosecond solid-state lasers at wavelengths between 1030 nm and 1064 nm (e.g., Nd: YAG and Ti: Saphir) with the possibility of higher harmonic generation in visible (515 nm–535 nm) and ultraviolet (UV) spectrum (342 nm–355 nm and 257 nm–266 nm), then copper vapor lasers, diode lasers, and excimer lasers emitting at UV region (157 nm–353 nm) [2].

Main process parameters in the laser-material interaction involve laser pulse duration. Consequently it significantly affects the quality of the produced microfeature and the material removal rate. Thermal relaxation time τ plays critical role during ablation. Thermal relaxation time is related to dissipation of heat during pulse irradiation and is expressed as

$$\tau = \frac{d^2}{\kappa},$$

(1)

where d is absorption depth and k is thermal diffusivity. Thus, high absorption coefficient α and low thermal diffusivity k ensure easy initiation of ablation [1].

Ablation of material can be facilitated by using short pulses (shorter than τ, generally 10 ps) as the laser energy is confined in a thin layer. For longer pulses, absorbed energy will be dissipated in the surrounding material by thermal processes. Absorption of long laser pulse also causes melting and substantial sputter evaporation of the material. These phenomena can contaminate surrounding area, produce microcracks, and remove material over dimensions much larger than the laser spot. Other adverse effects include damage to adjacent structures, delamination, formation of recast material, and formation of large heat affected zone (HAZ). Thus, efficient ablation of the material necessitates the use of ultrafast lasers. Ultrashort pulses produce very high peak intensity ($>10^{15}$ Wcm^{-2}) and deliver energy before thermal diffusion occurs. High efficiency and precision of the process are achieved without significant thermal degradation (melting, spatter, recrystallization, etc.) to the surrounding region. When such intense bursts of energy collide with the surface of any material, strong electromagnetic field of the focused laser pulse rips the electrons out of their atoms. Exposed material becomes ionized and thin plasma is created which results in material removal with extremely precise edges. The electrons are lighter and more energetic than the ions so they come off the material first, later followed by the ions. As the ions all have positive charge, they repel each other as they expand away from the material. Consequently, there are no droplets condensing onto the surrounding material [3].

EFFECT OF LASER PULSE DURATION

The interaction of laser radiation with matter was studied systematically. During the interaction absorbed laser energy is basically transferred to the electrons of machined material. The electrons reach thermal equilibrium in about 100 fs. After a certain period of time the electrons transmit their energy to the surrounding atoms or ions. This time interval is called electron-phonon relaxation time (τ) for materials with crystal structure and depends on the properties of the crystal lattice. Typical electron-phonon relaxation times differ from 0.5 ps to 50 ps. Thermal equilibrium between the electrons and the lattice is set in after a multiple of τ. Thus, no heat is transferred to the lattice for laser pulse durations shorter than τ. Therefore a melt phase or thermal damage in the material should not be present. Unfortunately situation in practice is different. As the lattice is heated up the evaporation takes place and persists for several nanoseconds. The material stays in molten state for tens of nanoseconds. So, even for ultrashort pulses, thermal processes last for time period in order of nanoseconds. Hence, the melt layer is never zero, but it reaches a thickness of several submicrons [1, 4].

LASER MICROMACHINING OF GLASS

Glass is suitable for many micro- and nanotechnology applications. Glass materials are characterized by attractive properties, such as inertness and other thermomechanical properties. Nowadays huge attention is being paid to glass machining because glass-based microstructures are found in disciplines such as biomedicine, biochemistry, lab-on-chip devices, sensors, and MEMS devices. Micro lenses and optical waveguides are key components for optical communication. In almost all cases, it is necessary to avoid damage and microcracking around laser machined site and femtosecond lasers have proven to be excellent sources for such precise machining work.

Generally, infrared (IR) laser radiation is not absorbed very well in variety of glass materials. It means that the band gap of the material is bigger than the photon energy and linear single photon absorption process cannot take place. If the material is exposed to high intensity femtosecond laser pulses the probability of nonlinear absorption mechanisms increases. For example, Li et al., 2008, fabricated microchannels and microchambers in silica glass and Darvishi et al., 2011, realized microchannels both in soda-lime and borosilicate glass using fs laser pulses [5, 6].

During micromachining, interaction of glass with laser beam is the most important part. Two types of phenomena can be observed when the beam is incident to glass, thermal expansion, and crystallization. Surface ablation occurs during both of these phenomena but during crystallization new phase is formed. To produce micro lenses for optical parts, the following conditions have to be kept: (1) no crystallization, because crystallization reduces the transmittance of glass and changes refractive index and (2) no contamination and cracking of the treated area. At this point glass has to satisfy high thermal stability and solidity [7].

To form the well-known waveguide structures it is desirable to focus the laser beam deeper under the glass surface. Then the local multiphoton interaction of laser pulses modifies refractive properties of glass creating waveguide structure. Laser fluence has to be greater than the surface damage threshold. This observation illustrates a significant difference of surface and internal laser-material interaction. In the material surface, the ionization stimulates avalanche processes and the resulting plasma plume either takes away material or causes a local breakdown with a following ejection of droplets. Internal multiphoton ionization most likely causes local photochemical reactions with a modified refractive index as a result [8].

Very thin glass sheets (with a thickness < 1 mm) are difficult to machine by conventional methods. These glass sheets are able to machine in situ with high quality, without damage and the formation of microcracks mainly by using ultrafast lasers [9, 10]. This method is confirmed by Yashkir and Liu, 2006, who presented ultrafast laser micromachining of 175 μm thick glass sheet using a Ti: Saphir laser. They demonstrated ultrafast microdrilling with HAZ limited to less than 1 μm and no presence of microcracks [8].

Methods using a laser with short wavelength or short pulse, glass with high absorptivity, and the addition of absorbents all have been proposed for the laser machining of glass without cracking. Laser radiation is absorbed at the interface of glass and the absorbent material and only a part of glass in contact with the absorbent is ablated. Mitsuishi et al., 2008, studied a method for machining a 3D microchannel in silica glass using a UV nanosecond pulsed laser and the slurry as absorbent. The machining specifications while using three different nanopowder materials, CeO_2, TiO_2, and ZnO, were investigated. Authors observed glass melting as a result of the heat transfer from absorbent particles, which were attached to the surface and assured strong laser absorption [11]. Kim et al., 2009, treated soda-lime and borosilicate glass doped with cobalt oxide as absorbent. Cobalt ions ensured that glass could be processed by laser. The irradiation area was smooth and no microcracks were present. The ablation height could be controlled by setting laser energy. Finally, type of glass used in this study was useful for the fabrication of microoptics parts [7].

One of the most promising indirect processing methods is the laser-induced back-side wet etching (LIBWE). During this method, transparent targets are in contact with liquid thin layers, which absorb and transform pulse energy resulting in etching. LIBWE is an effective method for crack-free etching of transparent materials such as glass achieving high precision and near-optical quality surfaces. Traditionally, LIBWE is performed using UV laser sources. However, Cheng et al., 2006, described the use of an economic Q-switched 532 nm green laser in microfabrication of soda-lime glass substrates. Using a common organic dye (Rose Bengal) as the photoetchant, crack-free microstructures with a minimum feature size of 18 µm were obtained [12]. Yen et al., 2010, used visible LIBWE with gallium and eutectic indium/gallium as absorbers, for crack-free microfabrication of soda-lime and quartz glass [13]. Ultraviolet surface microstructuring of silica glass plates by LIBWE was performed upon irradiation from solid state UV laser (266 nm) by Niino et al., 2006. They reported a well-defined micropattern formation without debris and microcrack around the etched area. Zimmer and Böhme, 2008, suggested hydrocarbon and metallic absorbers for LIBWE of transparent materials [14, 15].

Demand for higher precision and clean laser based processes has driven the development of the ultrafast lasers that operate at high frequency with high average powers. Lee et al., 2009, focused on the differences between processing with nanosecond and picosecond pulses on aluminoborosilicate glass. Comparison of trepanned holes produced in glass using nanosecond and picosecond pulses revealed that the wall surfaces and the entrance holes produced by picosecond pulses were smoother than those by nanosecond ones. The picosecond laser provided high quality processing at lower speeds compared to nanosecond [16]. Although Ramil et al., 2008, demonstrated the feasibility of micromachining of soda-lime glass using the third harmonic of nanosecond Nd: YVO_4 laser [17], laser cutting of glass was presented by Loeschner et al., 2008. The investigations were carried out with a short nanosecond pulsed Nd: YVO_4 slab laser and a high repetition rate femtosecond laser. Irradiation of the material with short nanosecond pulses leads to the formation of micro defects [10].

LASER MICROMACHINING OF SILICON

Silicon is one of the most investigated materials. It is widely used in semiconductor industry and it is very often exploited as a substrate for many electronic devices. Polycrystalline and amorphous silicon is suitable for solar cell technology and in various chips. Many studies have compared fs- and ns-laser technology and it was established that short pulses, shorter than 15 ns and wavelength in UV region, reduce thermal effects (HAZ or deposition of molten material). Picosecond lasers have also found a number of applications in solar cells and other silicon-based applications [16, 18].

The precision of laser microfabrication is influenced by absorption, mechanism of heat transport, and energy density of the laser radiation. Figure 1 shows absorption of silicon as a function of wavelength at 25°C for a 600 μm thick sample. The graph demonstrated that the radiation with wavelength below about 380 nm is absorbed by silicon significantly better. As the temperature rises, it is expected that absorption will increase as the material becomes molten [18].

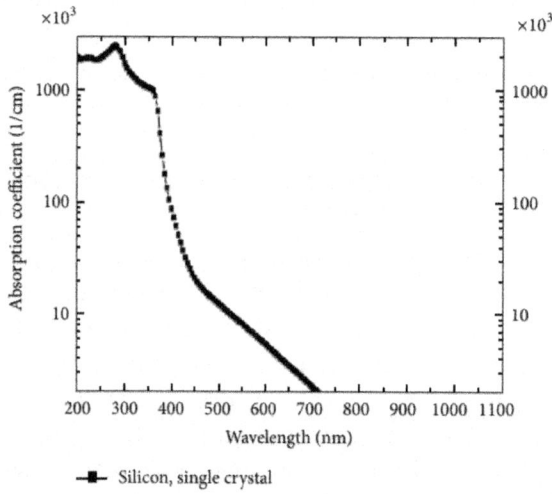

Figure 1: Absorption coefficient of silicon depending on the wavelength at 25°C for a 600 μm thick polished sample [18].

Klotzbach et al., 2011, observed creation of 15 μm wide craters surrounded by 40 μm wide area of splashed melt after the application of one pulse of radiation of Nd: YAG laser emitting at 355 nm. They also focused on percussion drilling of silicon and drilled holes with diameter of 25 μm. Silicon cutting was performed by Nd: YAG lasers emitting at 355 nm and 532 nm. They found that cutting with shorter wavelength reduced the formation of melt phase and thus improved the quality of the cut [18]. Single-shot laser ablation of monocrystalline silicon with a nanosecond Nd: YAG laser at 355 nm and energy density from 10^9 Jcm^{-2} to 10^{11} Jcm^{-2} was also investigated by Karnakis, 2005 [19]. Formed craters were measured and their morphologies were analyzed. The study reported that the depth of craters increased with the increasing power density. Significant laser induced melting and droplets were presented around crater rims.

Ablation of silicon was investigated using a 1064 nm pulsed fiber laser, with pulse energy up to 0.5 mJ, peak powers up to 10 kW, and pulse widths from 10 ns to 250 ns by Hendow and Shakir, 2010. They indicated that pulses with high peak powers caused the decrease of penetration depths, while longer pulses, with lower peak powers, had a higher material removal rate with deeper scribes as ablation was accompanied by preheating of the surface, melting of the material, and a shock wave that ejected the molten material [20]. Herbst et al., 2001, investigated machining silicon wafers with thickness up to 1 mm using a diode-pumped-solid-state laser that delivered short pulses of about 15 ns at 355 nm wavelength. The results showed a small HAZ with little evidence of microcracking. But even a small increase in laser pulse width could increase the thermal damage of silicon that could be seen as an enlarged amount of the redeposited material in and around the hole [21].

Lee et al., 2009, also focused on silicon processing and optically compared drilled holes created by picosecond and nanosecond UV laser systems. The resulting hole formed by picosecond laser was cleaner on the surface and the internal wall was smoother. The nanosecond hole had some oxide residue on the top and on the wall surface. The picosecond laser provided high quality processing at lower speeds comparing to nanosecond [16]. No chipping and HAZ along the cut produced by picosecond emitting at 515 nm were confirmed also by Weiler et al., 2008 [22]. Bärsch et al., 2003, Karnakis et al., 2005, and Rizvi et al., 2011, reported that ultrashort pulses are the best choice for machining silicon as it provides excellent results, high quality, precision, and no damage [23–25]. Ren et al., 2005, and Kruusing, 2004, realized ablation of silicon under water. Their results showed that ablation in the water is faster than in the air and is characterized by small HAZ and no mechanical damage [26, 27].

LASER MICROMACHINING OF CERAMICS

A variety of ceramic materials such as zirconium and aluminum carbide are widely used in the field of microelectronics and in other MEMS-type devices. The test devices for integrated circuits, substrates for sensors and detectors, microcavity structures inside biomedical or chemical diagnostics, and transducers are possible applications.

Several types of lasers, for example, CO_2, Nd: YAG, and excimer lasers, are used for machining of structural ceramics. Pulsed lasers are preferred for machining ceramics due to better and more effective control of process parameters compared to continuous wave mode. Four physical phenomena can be distinguished when the laser beam is incident on the ceramic surface. These are reflection, absorption, scattering, and transmission. Absorption, crucial of all the phenomena, is described as the interaction of electromagnetic radiation with the electrons of the material. It depends on both the wavelength and the spectral absorptivity characteristics of ceramics being machined (e.g., reflection coefficient). The absorptivity is also influenced by the orientation of the ceramic surface with respect to the beam direction and reaches a maximum value for angles of incidence above 80°. Thermal conductivity of structural ceramics is generally smaller than that of most metals; so radiation absorption sets in faster. Absorbed energy is converted into heat and its subsequent conduction into the material establishes the temperature distribution within the material that in turn affects machining time and depth of cavity [28].

Ablation occurs when laser energy exceeds the characteristic threshold that represents the minimum energy required to remove the material by ablation. Laser machining of ceramics is quite hard because of large scattering that appears for many common laser wavelengths, which restricts localized energy absorption. The ablation threshold is higher for metals (factor of 2–10) and more clearly defined. A combination of short pulses and short wavelengths usually leads to the best results and in many ceramics, such as alumina and silicon nitride melting, is not evident. Thermal stresses during micromachining can result in cracking, so that optimized processing parameters should be set to keep the heat input to the bulk material low and thus to avoid cracks formation. Mostly, this is ensured by precluding the formation of intense laser induced plasma. Micromachining with longer pulse lasers (micro- and millisecond regime) includes melting in material removal mechanism. This method promises processing with very high removal rates but causes a creation of glassy layer that is often regarded as the source of microcracks [4].

Nedialkov et al., 2003, studied laser ablation of alumina, aluminum nitride, and silicon nitride using nanosecond Nd: YAG laser with different wavelengths. They found that IR radiation provided the highest ablation rate. Drilling holes in Si_3N_4 had the best quality in respect to the debris and roundness compared to other ceramics [29]. Liu et al., 2007, presented ultrafast structuring of ceramics using femtosecond laser, whereas Karnakis et al., 2006, used nanosecond copper vapor laser (511 nm) and picosecond Nd: YVO_4 (1064 nm) to drill holes in ceramic materials. They reported that both nanosecond and picosecond lasers can be used for high quality laser micromilling of ceramics that is difficult to machine with ultrahigh precision using conventional methods. Both laser types were capable of excellent surface finish with relatively high material removal rates [30, 31].

Liu et al., 2007, confirmed precise, melt-free ultrafast laser microstructuring of ceramic alumina. Wang et al., 2010, also showed the results of femtosecond laser drilling of alumina ceramic substrate. They investigated the effects of various laser parameters such as different focus position, traverse speed, drilling pattern, and pausing time on the drilled holes quality, HAZ, holes circularity, or debris production. They demonstrated high-quality laser drilling with clean surface, no cracks, no recast layer, and no delamination which can be used in manufacturing of electronic devices [32]. Ho et al., 2010 created blind and through micro holes by percussion drilling as well as trepanning drilling using picosecond Nd: YLF laser. The diameters of the holes were in the range of 20 µm–1000 µm [33]. Perrie et al., 2005, and Kim et al., 2008 and 2009, realized microfabrication of alumina and nitride ceramics using femtosecond lasers. Perrie et al., 2005, compared femtosecond microstructuring of alumina ceramics with nanosecond UV processing. Femtosecond pulses caused excellent edge quality and no discoloration of the treated surface, unlike those produced by nanosecond UV pulses. Kim et al., 2008, drilled micro holes with sub-100 µm diameter, whereas Kim et al., 2009, investigated aluminum oxide and aluminum nitride ablation characteristics, specifically the threshold fluence, incubation effect, and ablation rate. The ablation characteristics of the two ceramics showed similar trends except for surface morphologies, which revealed virtually no melting in Al_2O_3 but clear evidence of melting for AlN [34–36]. Bärsch et al., 2007, studied microstructuring of zirconium ceramics by femtosecond laser, whereas Zeng et al., 2007, used nanosecond Nd: YAG laser [37, 38].

CONCLUSION

In this paper we report the mechanisms of laser micromachining of materials. It is very promising technique in many industries. It is possible to machine various materials using lasers with different pulse length. In most cases ultrafast lasers bring better resolution than lasers with longer pulses. We concluded that ultrafast micromachining is characterized by excellent quality of treated materials, high precision of the process, small HAZ, and production of no microcracks.

ACKNOWLEDGMENT

The authors gratefully acknowledge the support of the project Coherent and Nonlinear Optics, Selected Chapters V IGA_PrF_2014005.

REFERENCES

1.   N. B. Dahotre and S. P. Harimkar, Laser Fabrication and Machining of Materials, Springer, New York, NY, USA, 2008.

2.   D. Karnakis, Ultrafast Laser Nanomachining: Doing More With Less, Commercial MicroManufacturing, Oxford, 2008.

3.   F. Dausinger, H. Hügel, and V. I. Konov, "Micromachining with ultrashort laser pulses: from basic understanding to technical applications," in International Conference on Advanced Laser Technologies (ALT '02), vol. 5147 of Proceedings of SPIE, p. 106, November 2003.

4.   M. R. H. Knowles, G. Rutterford, D. Karnakis, and A. Ferguson, "Micro-machining of metals, ceramics and polymers using nanosecond lasers," The International Journal of Advanced Manufacturing Technology, vol. 33, no. 1-2, pp. 95–102, 2007.

5.   Y. Li, D. Liu, F. Qi, H. Yang, and Q. Gong, "Femtosecond laser micromachining and microfabrication in transparent materials," in Lasers in Material Processing and Manufacturing III, vol. 6825 of Proceedings of SPIE, pp. 1–10, February 2008.

6.   S. Darvishi, T. Cubaud, and J. P. Longtin, "Ultrafast laser machining of tapered microchannels in glass and PDMS," Optics and Lasers in Engineering, vol. 50, no. 2, pp. 210–214, 2012.

7.   T. H. Kim, Y. S. Kim, Y. J. Jeong et al., "Micromachining of transition metal ion doped glass surface by using a pulsed Nd:YAG (532 nm) laser for the optical device," Current Applied Physics, vol. 9, no. 3, pp. 234–236, 2009.

8.   Y. Yashkir and Q. Liu, "Experimental and theoretical study of the laser micromachining of glass using a high-repetition-rate ultrafast laser," in Solid State Lasers and Amplifiers II, vol. 6190 of Proceedings of SPIE, pp. 236–245, Strasbourg, France, April 2006.

9.   N. H. Rizvi, "Femtosecond laser micromachining: current status and applications," Riken Review, vol. 50, pp. 107–112, 2003.

10. U. Loeschner, S. Mauersberger, R. Ebert et al., "Micromachining of glass with short ns-pulses and highly repetitive fs-laser pulses," in Proceedings of the 27th International Congress on Applications of Lasers and Electro-Optics (ICALEO '08), pp. 193–201, October 2008.

11. M. Mitsuishi, N. Sugita, I. Kono, and S. Warisawa, "Analysis of laser micromachining in silica glass with an absorbent slurry," CIRP Annals—Manufacturing Technology, vol. 57, no. 1, pp. 217–222, 2008.

12. J.-Y. Cheng, M.-H. Yen, and T.-H. Young, "Crack-free micromachining

on glass using an economic Q-switched 532 nm laser," Journal of Micromechanics and Microengineering, vol. 16, no. 11, article 24, 2006.

13. M.-H. Yen, C.-W. Huang, W.-C. Hsu, T.-H. Young, K. Zimmer, and J.-Y. Cheng, "Crack-free micromachining on glass substrates by visible LIBWE using liquid metallic absorbers," Applied Surface Science, vol. 257, no. 1, pp. 87–92, 2010.

14. H. Niino, Y. Kawaguchi, T. Sato, A. Narazaki, and R. Kurosaki, "Surface microfabrication of silica glass by LIBWE using DPSS-UV laser," in Photon Processing in Microelectronics and Photonics V, 61061E, vol. 6106 of Proceedings of SPIE, San Jose, Calif, USA, January 2006.

15. K. Zimmer and R. Böhme, "Laser-induced backside wet etching of transparent materials with organic and metallic absorbers," Laser Chemistry, vol. 2008, Article ID 170632, 13 pages, 2008.

16. S. Lee, A. Ashmead, and L. Migliore, "Comparison of ns and ps pulses for Si and glass micromachining applications," in Solid State Lasers XVIII: Technology and Devices, vol. 7193 of Proceedings of SPIE, p. 10.

17. A. Ramil, J. Lamas, J. C. Álvarez, A. J. López, E. Saavedra, and A. Yáñez, "Micromachining of glass by the third harmonic of nanosecond Nd:YVO$_4$ laser," Applied Surface Science, vol. 255, no. 10, pp. 5557–5560, 2009.

18. F. A. Lasagni and A. F. Lasagni, Fabrication and Characterization in the Micro-Nano Range, Springer, 2011.

19. D. M. Karnakis, "High power single-shot laser ablation of silicon with nanosecond 355 nm," Applied Surface Science, vol. 252, no. 22, pp. 7823–7825, 2006.

20. S. T. Hendow and S. A. Shakir, "Structuring materials with nanosecond laser pulses," Optics Express, vol. 18, no. 10, pp. 10188–10199, 2010.

21. L. Herbst, J. P. Quitter, G. M. Ray et al., "High peak power solid state laser for micromachining of hard materials," in Solid State Lasers XII, vol. 4968 of Proceedings of SPIE, pp. 134–142, San Jose, Calif, USA, June 2003.

22. S. Weiler, U. Stute, S. Massa, S. Buettner, and B. Faisst, "Efficient micro machining with high average power picosecond lasers," in Commercial and Biomedical Applications of Ultrafast Lasers VIII, vol. 6881 of Proceedings of SPIE, January 2008.

23. N. Bärsch, K. Körber, A. Ostendorf, and K. H. Tönshoff, "Ablation and cutting of planar silicon devices using femtosecond laser pulses," Applied Physics A, vol. 77, no. 2, pp. 237–242, 2003.

24. D. M. Karnakis, G. Rutterford, and M. R. H. Knowles, "High power DPSS laser micromachining of silicon and stainless steel," in Proceedings of the Lasers in Manufacturing—WLT Conference, pp. 741–746, 2005.

25. N. H. Rizvi, D. Karnakis, and M. C. Gower, "Micromachining of industrial materials with ultrafast lasers," in Proceedings of the 20th International Congress on Applications of Lasers & Electro-Optics (ICALEO ‹01), pp. 1511–1520, October 2001.

26. J. Ren, M. Kelly, and L. Hesselink, "Laser ablation of silicon in water with nanosecond and femtosecond pulses," Optics Letters, vol. 30, no. 13, pp. 1740–1742, 2005.

27. A. Kruusing, "Underwater and water-assisted laser processing: part 2—etching, cutting and rarely used methods," Optics and Lasers in Engineering, vol. 41, no. 2, pp. 329–352, 2004.

28. A. N. Samant and N. B. Dahotre, "Laser machining of structural ceramics—a review," Journal of the European Ceramic Society, vol. 29, no. 6, pp. 969–993, 2009.

29. N. N. Nedialkov, P. A. Atanasov, M. Sawczak, and G. Sliwinski, "Ablation of ceramics with ultraviolet, visible, and infrared nanosecond laser pulses," in XIV International Symposium on Gas Flow, Chemical Lasers, and High-Power Lasers, vol. 5120 of Proceedings of SPIE, pp. 703–708, November 2003.

30. D. Karnakis, G. Rutterford, M. Knowles, T. Dobrev, P. Petkov, and S. Dimov, "High quality laser milling of ceramics, dielectrics and metals using nanosecond and picosecond lasers," in 5th Photon Processing in Microelectronics and Photonics, vol. 6106 of Proceedings of SPIE, San Jose, Calif, USA, March 2006.

31. D. Liu, J. Cheng, W. Perrie et al., "Femtosecond laser micro-structuring of materials in the NIR and UV regime," in Proceedings of the 26th International Congress on Applications of Lasers and Electro-Optics (ICALEO ‹07), pp. 12–18, November 2007.

32. X. C. Wang, H. Y. Zheng, P. L. Chu et al., "Femtosecond laser drilling of alumina ceramic substrates,"Applied Physics A: Materials Science and Processing, vol. 101, no. 2, pp. 271–278, 2010.

33. C.-Y. Ho, Y.-H. Tsai, Ch.-S. Chen, and M.-Y. Wen, "Ablation of aluminum oxide ceramics using femtosecond laser with multiple pulses," Current Applied Physics, vol. 11, no. 3, supplement, pp. S301–S305, 2011.

34. W. Perrie, A. Rushton, M. Gill, P. Fox, and W. O›Neill, "Femtosecond laser micro-structuring of alumina ceramic," Applied Surface Science, vol. 248, no. 1–4, pp. 213–217, 2005.

35. S. H. Kim, T. Balasubramani, I. B. Sohn et al., "Precision microfabrication of AlN and Al_2O_3 ceramics by femtosecond laser ablation," in Photon Processing in Microelectronics and Photonics VII, vol. 6879 ofProceedings of SPIE, January 2008.

36. S. H. Kim, I.-B. Sohn, and S. Jeong, "Ablation characteristics of aluminum oxide and nitride ceramics during femtosecond laser micromachining," Applied Surface Science, vol. 255, no. 24, pp. 9717–9720, 2009.

37. N. Bärsch, K. Werelius, S. Barcikowski, F. Liebana, U. Stute, and A. Ostendorf, "Femtosecond laser microstructuring of hot-isostatically pressed zirconia ceramic," Journal of Laser Applications, vol. 19, no. 2, pp. 107–115, 2007.

38. D. W. Zeng, K. Li, K. C. Yung, H. L. W. Chan, C. L. Choy, and C. S. Xie, "UV laser micromachining of piezoelectric ceramic using a pulsed Nd:YAG laser," Applied Physics A: Materials Science and Processing, vol. 78, no. 3, pp. 415–421, 2004.

Chapter 7

MODELING AND SIMULATION OF PROCESS-MACHINE INTERACTION IN GRINDING OF CEMENTED CARBIDE INDEXABLE INSERTS

Wei Feng[1] Bin Yao[1] BinQiang Chen[2] DongSheng Zhang[3] XiangLei Zhang[1] and ZhiHuang Shen[1]

[1]Department of Mechanical and Electrical Engineering, School of Physics and Mechanical & Electrical Engineering, Xiamen University, Xiamen 361005, China

[2]State Key Laboratory for Manufacturing and Systems Engineering, School of Mechanical Engineering, Xi'an Jiaotong University, Xi'an 710049, China

[3]School of Mechanical Engineering, Shaanxi University of Technology, Hanzhong 723001, China

ABSTRACT

Interaction of process and machine in grinding of hard and brittle materials such as cemented carbide may cause dynamic instability of the machining process resulting in machining errors and a decrease in productivity. Commonly, the process and machine tools were dealt with separately, which does not take into consideration the mutual interaction between the two subsystems and thus cannot represent the real cutting operations. This paper proposes a method of modeling and simulation to understand well the process-machine interaction in grinding process of cemented carbide indexable inserts. First, a virtual grinding wheel model is built by considering the random nature of abrasive grains and a kinematic-geometrical simulation is adopted to describe the grinding process. Then, a wheel-spindle model is simulated by means of the finite element method to represent the machine structure. The characteristic equation of the closed-loop dynamic grinding system is derived to provide a mathematic description of the process-machine interaction. Furthermore, a coupling simulation of grinding wheel-spindle deformations and grinding process force by combining both the process and machine model is developed to investigate the interaction between process and machine. This paper provides an integrated grinding model combining the machine and process models, which can be used to predict process-machine interactions in grinding process.

INTRODUCTION

High performance cutting tools play an important role in modern manufacturing. To some extent, the performance of a cutting tool determines the cutting behavior and the process capability. With recent developments in aerospace and automobile industries, there are increasing demands for high-quality parts [1,2]. In order to meet the urgent requirements for high quality parts, cutting tools made of cemented carbide are used, which is a very hard and brittle material. Due to the hardness of cemented carbide cutting tools are mainly manufactured by grinding process. Grinding is one of the most common material removal processes to achieve desired surface integrity, dimensional tolerance, and form tolerance. It is used as a typical finishing process as well as for high material removal rates. In the grinding process, the machine tools and process constitute a closed loop machining system. The generated cutting forces and temperature cause elastic deformations in the machine tool system which change the instantaneous chip area which in turn has an influence on the cutting forces.

During precision machining, the dynamic interactions between machining process and grinding machine structure, for example, vibrations, deflections, or thermal deformations, result in poor quality of produced parts, short life of machine components and tools [3, 4]. Different methods have been developed and used to detect and diagnose the machine faults [5–7]. In the past, the process and machine tools were dealt with separately. However, the separation of the machining system into the two subsystems does not take into consideration the mutual interaction between the two subsystems and thus cannot represent the real cutting operations. The continuous and mutual influence exerted by both machine and process results in the often unpredictable effects in precision grinding of hard and brittle materials such as cemented carbide, which may cause dynamic instability of the machining process resulting in machining errors and a decrease in productivity.

In order to meet requirements of high accuracy and productivity in grinding of cemented carbide cutting tools, it is essential to understand well the interaction between the process and the machine tool system linked with the force and the deformation. The interaction between process and machine is usually hard to predict due to its complexity, and extensive experiments are laborious and expensive. In many cases, predictions can only be made by means of complex simulations [3]. There has been numerous researches conducted to the modeling and simulation of grinding processes in the last two decades; however, commonly most focus on simulation of local contact area interaction between the grinding wheel and workpiece and neglect displacements of the machine system [8–12]. Once vibrations and deformations of machine structure caused by process forces can be determined, strategies can be derived to reduce geometry as well as thermomechanical errors by process parameter and tool path optimization [4]. Therefore, it is necessary to treat process and machine structure in an integrated way.

The prediction of process-machine interactions requires models of machine and process. To predict the interactions these models have to be built and then coupled and simulated. Due to the large number of abrasive grains with unknown time-dependent geometry and distribution, grinding is a complex material-removal process [13]. Based on kinematic-geometrical simulations, numerous macroscopic and microscopic approaches have been used to build the grinding process model up to now [9]. On the machine tool side, the integrated models providing data on static and dynamic behavior of the structure must be appropriately parameterized. Multiple approaches for building a machine model are presented in [10].

The simulation of cutting processes under consideration of process and machine properties can be carried out along with different methods. The coupled simulation was a relatively new approach which permits simultaneous usage of two different simulation environments with data exchange by means of a suitable interface [3]. Brecher and Witt [14] proposed an approach for simulating a machining process, including interactions between machine tool and a cutting process. Three different approaches were used to carry out a coupled simulation of machine and process. Herzenstiel et al. [15] developed a comprehensive simulation of the grinding process and the grinding machine. They combined the process model (KSIM) and the machine model (FEM) and proposed a staggered iteration scheme. Weinert et al. [16] used a cosimulation comprising a geometric-kinematical process simulation and a finite elements simulation to investigate NC shape grinding. The authors also provided an iterative solution for nonlinear process-structure coupling problems. Aurich and Kirsch [17] presented new insights in the modeling process and the examination of process-machine interactions using kinematic simulation.

In this paper, the interaction of process and machine in cemented carbide insert grinding is studied. The authors propose a geometric-kinematical simulation and a wheel-spindle model based on finite element method. Furthermore, a couple simulation approach which treats the grinding process and machine tool structure in an integrated manner is presented. Also, an investigation of the results of process-machine interaction is described.

PROCESS MODEL

A kinematic-geometrical simulation model (KSIM) is adopted in this research work, which was first developed by Warnecke and Zitt [18]. The concept of

KSIM is based on the observed micro- and macroscopic cause and effect chain in grinding and simulates the grinding process as a penetration between the enveloping profile of the grinding wheel and the workpiece [17]. Therefore, the micro and macro geometry of the grinding wheel and workpiece have to be modeled. In the simulation, ideal cutting is assumed by not taking account of rubbing and ploughing of materials and each grain removes the whole material volume encountered. In the simulation, the loads and stresses can be calculated in FEM for every single grain by using material models. On the other hand, the forces that are generated can be calculated using modified Kienzle equations together with the knowledge of the accumulated machined material and the chip corss-section [3].

Modeling of the Grinding Wheel

To gain realistic simulation results the grinding wheel topography has to be modeled precisely. The topography of a diamond wheel in a stationary wear state was measured by using a KEYENCE VHX-5000 3D-measuring system. Figure 1 shows the top view of the topography at working face. As can be seen from Figure1, the real geometry of single grain is similar to pyramid; therefore hexahedron was chosen in this work for mathematical simplification in modeling. It is noted that the process model to be built requires considering the combined action of the grains which are stochastically spaced in the grinding wheel. The authors proposed a novel approach to take account for the random nature of grinding wheel [19]. The average grit spacing is estimated by using the density of the grits. The location of each grain is represented by its center coordinate. The spatial probability distribution of the grains is obtained by adjusting the location of each individual grain center in three-dimensional space.

Figure 1: Topographies of grinding wheel.

In order to avoid the overlapping between abrasive grains in the binder, a virtual grid approach is used to account for the random nature of the grinding wheel. Assuming abrasive grains are distributed in a square area of the grinding wheel and each grain is restrained in an imaginary grid, see Figure 2. A is the length of the virtual grid which depends on the grain density. D is the diameter of the grain. The location coordinates of the grains are randomly distributed in a $l \times l$ square area.

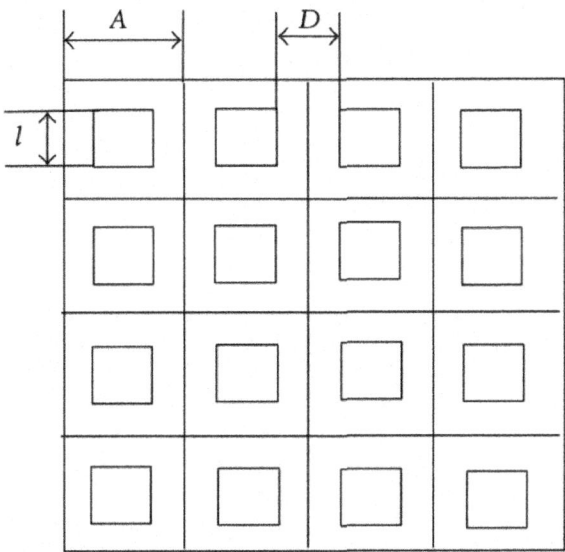

Figure 2: Virtual grids of abrasive grains distribution.

The major components of the topography of grinding wheel are abrasive grains and bonds. Since only the grains participate in actual cutting process, the influence of binds is neglected in modeling. For modeling the grinding, a real diamond 12A2T grinding wheel in stationary wear state with a dimension of Φ400 × 4.8 mm was used as archetype; see Figure 3. The density of the grits was 127 per mm² and the nominal grit size was between 50 μm and 70 μm. As to the grain geometry, a hexahedron with each side length of 50 μm was adopted. A method in [19] was used to build the detail grinding wheel model. First, the random position coordinates of the abrasive grains were generated; see Figure 4. Then, the geometry model was built with a commercial three-dimensional software. For a section of the modeled virtual grinding wheel, see Figure 5.

Figure 3: Virtual grinding wheel archetype.

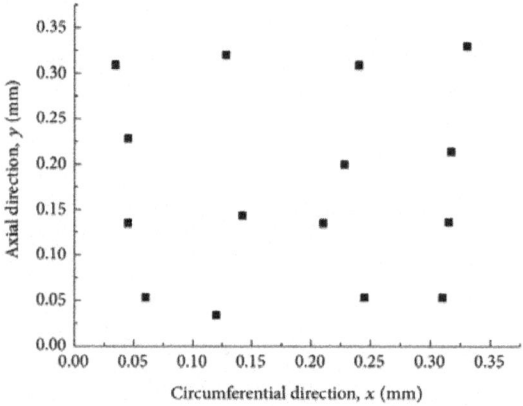

Figure 4: Random position coordinates of grains.

Figure 5: Modeled grinding wheel section.

Process Model

With the grinding wheel model at hand, simulations for different combinations of the input grinding parameters can be conducted. In the simulation, ideal cutting is assumed neglecting ploughing and friction effects; therefore the process force at a given time depends linearly on the cutting depth based on Kienzle approach:

$$F(t) = k_c \cdot \Delta(t),$$

(1)

where k_c denotes the process stiffness. The output parameters of the simulation are the experimentally not measurable undeformed chip thickness, chip length, chip width and chip cross section of every single grain. Using the simulated chip cross section from simulation, the process force can be calculated. Process stiffness k_c is an experimentally determined parameter, which can be determined by executing series of tests. In this work, the face grinding of a cemented carbide indexable insert is investigated; see Figure 6. The details of the grinding tests can be found in [19]. It has to be noted that the grinding tests conducted in this work were based on a fixed y-axis, which was different from [19].

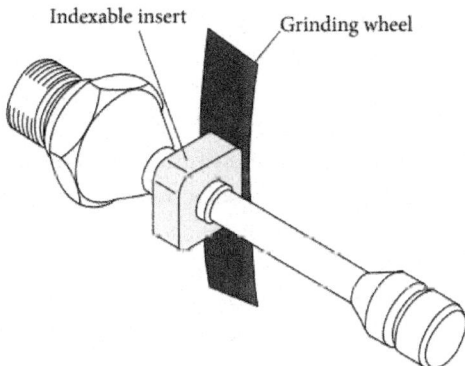

Indexable insert   Grinding wheel

Figure 6: Cemented carbide cutting insert grinding.

The exemplary parameter combination of cutting speed 42 m/s, depth of cut 0.02 mm, and feed rate 300 mm/min was chosen to show and explain the results of simulation. In Figure 7 the grinding force for one wheel revolution of the simulated process force is depicted. Due to the constant change of protrusion heights on the topography of grinding wheel, the chip cross-sectional areas fluctuate and result in an oscillating simulated force signal.

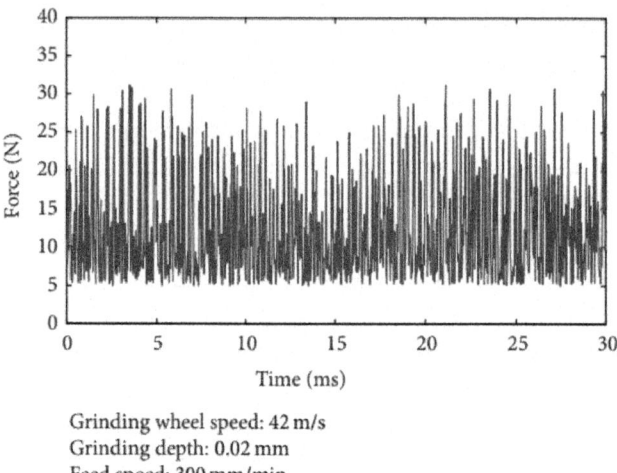

Grinding wheel speed: 42 m/s
Grinding depth: 0.02 mm
Feed speed: 300 mm/min

Figure 7: Simulated process force.

MACHINE MODEL

For an exact simulation of the mutual interaction between grinding process and machine, the machine structure has to be modeled accurately. Therefore the machine model has to be able to respond to excitations coming from the grinding process and to create an output which again has an influence on the calculated grinding process. Since the modeling of a complete machine is usually very complex and time consuming, if some of the machine components react to an excitation by the grinding process, a selective model can be adopted [4]. The main modeling methods for grinding machines are finite element, multibody simulation, boundary element, and analytical methods [3]. In this work, a wheel-spindle finite element (FE) model is chosen and built; see Figure 8. The spindle is made of steel and the wheel hub is made of 40Cr. Material properties are listed in Table 1. Apart from the wheel-spindle model, the contact between wheel and workpiece can be simplified to single point contact [12]; see Figure 9. Accordingly, only one node is in contact. The contact force returned for a single contact point is a function of time and displacements.

Table 1: Material properties of spindle and wheel hub

	Steel	40Cr
Density [kg/m^3]	7890	7870
Young modulus [GPa]	209	211
Poisson's ratio	0.269	0.277

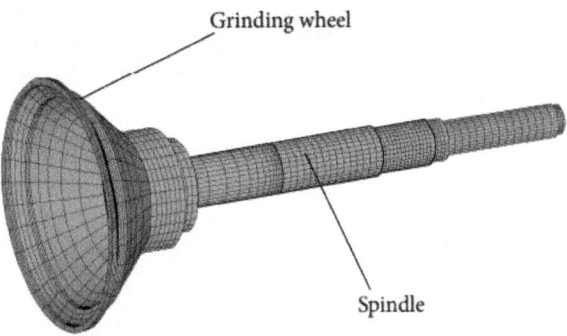

Figure 8: Wheel-spindle FE model.

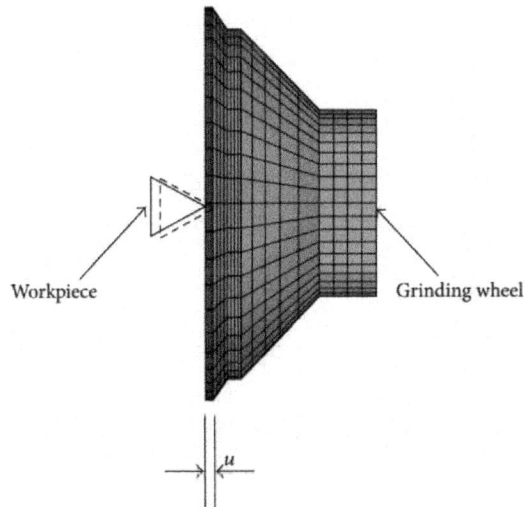

Figure 9: Point contact.

For parameterising the wheel-spindle model, static and dynamic behavior of the real machine must be measured and characterized accurately. A modal experiment was conducted by using an impact hammer instrumented with a piezoelectric transducer and an accelerometer. Then the measured eigenfrequencies were compared with the results of the modal analysis by using commercial FEA software to verify and validate the related FE model. All eigenfrequencies extracted from FE simulations are well matched to the experimental data; see Table 2.

Table 2: Eigenmodes 1–5 extracted from experimental data and FE simulations

Mode	Experiment [Hz]	FE model [Hz]	Difference [%]
1	220	215.62	1.99%
2	330	353.64	7.16%
3	780	753.23	3.43%
4	950	929.53	2.16%
5	1050	1019	2.95%

PROCESS-MACHINE INTERACTION

Modeling of Process-Machine Interaction

Assuming the dynamic displacements of machining system under excitation of grinding process forces at a given time will be

$$x(t) = \Delta(t) - \Delta(t - T),$$

$$(2)$$

where $\Delta(t)$ and $\Delta(t-T)$ are the present and past instantaneous grinding depth, T is grinding rotation period. Assuming the nature frequency of the machine tool is w, the variation of dynamic displacement (t) in the frequency domain can be represented as

$$X(iw) = G_m(iw) F(iw),$$

$$(3)$$

where $G_m(iw)$ is the transfer function of the machine tool. Convert (1) and (2) from the time domain to the frequency domain and combine with (3); we can obtain the characteristic equation of the closed-loop dynamic grinding system

$$\left[1 - \left(1 - e^{-iwT}\right) k_c G_m(iw)\right] \Delta(iw) = 0.$$

$$(4)$$

With (4) the dynamic behavior of machine tool $G(iw)$ and the grinding parameters k_c, Δ are linked. The dynamic forces generated in grinding process continually excite the machine structure. The local cutting depth of grinding wheel varies, which in turn influences the grinding force. When the exciting frequency approaches the nature frequency of the machine structure, instability of the grinding process occurs.

Coupling Simulation

In order to simulate the interaction of grinding process and machine structure, a coupling simulation method is adopted in this work. In coupling simulation, both simulations run simultaneously in different simulation environments and communicated to each another in synchronized cycles [3]; see Figure 10.

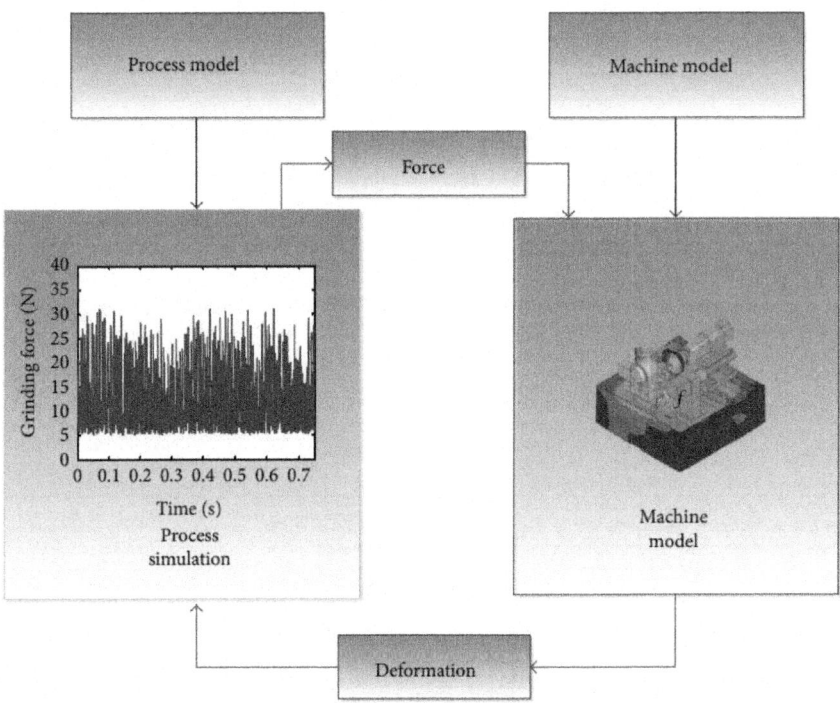

Figure 10: Coupling simulation of process-machine interaction.

A simulation cycle is developed in a coupling simulation. First, a simulated force signal generated by the process model starts as input data for the machine model. The machine model delivers displacements related to the contact zone for one wheel revolution, which again constitutes input data for the process model. Equation (1) is only valid for an ideally stiff grinding wheel; however, grinding wheel has axial deformation u due to forces in grinding process; see Figure 9. Assuming the initial grinding depth is Δ_0, the actual grinding depth will be $(\Delta_0 + u)$. By multiplying $(\Delta_0 + u)$ by the process stiffness kc an adjusted force value is then obtained and can be subsequently used to calculate the deformation again. Since the process and machine state vary in time, the coupling simulation has to be solved iteratively. Weinert et al. proposed an iteration approach to solve the problem [16]. In order to evaluate the process and machine state after each iteration step, the convergence criteria have to be given. The iteration is repeated until the simulated force or displacement signals of two consecutive iteration steps show minor deviations; see Figure 11. Figure 12 shows the mean values for force and the displacement converge over one grinding wheel revolution after 5 iterations from (31.9 N, 11.56 μm) to (29.2 N, 8.5 μm). These values are smaller than the values corresponding to the uncoupled solution.

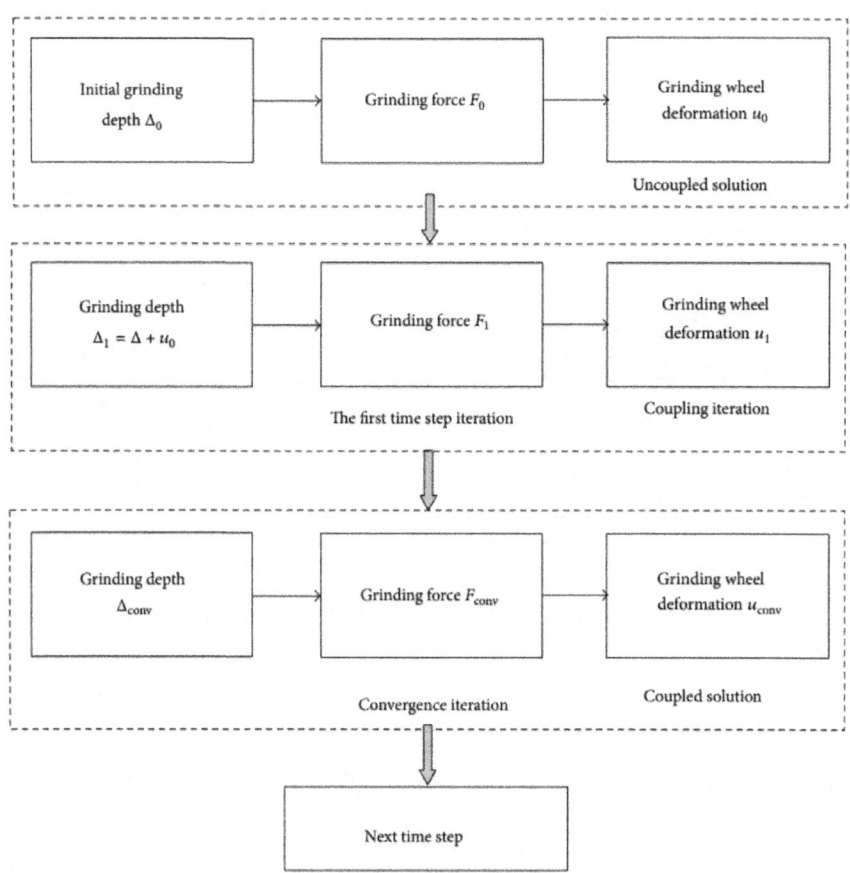

Figure 11: Data exchange and convergence for iteration process.

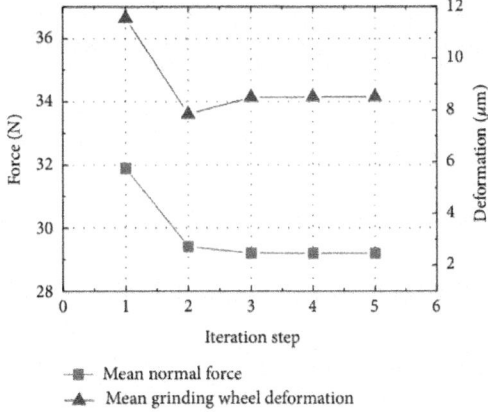

Figure 12: Iteration convergence behavior.

Coupling Results

A fast Fourier transform of the force signal in Figure 6 shows the exciting frequencies of grinding forces with a dominating contribution of frequencies of up to 1100 Hz; see Figure 13. As can be seen from Figure 13, the dominating frequencies of force signal cover the top 5 order frequencies of wheel-spindle structure. The force signal can be substituted by a set of harmonic function of harmonic depths of cut and the generated harmonic loads are able to excite the grinding wheel. The force and the deformation are normalized so that they can be represented in the same diagram [20]. All forces shown are normalized by dividing the amplitudes of the initial force value. And all deformations are normalized by dividing the maximum deformation value.

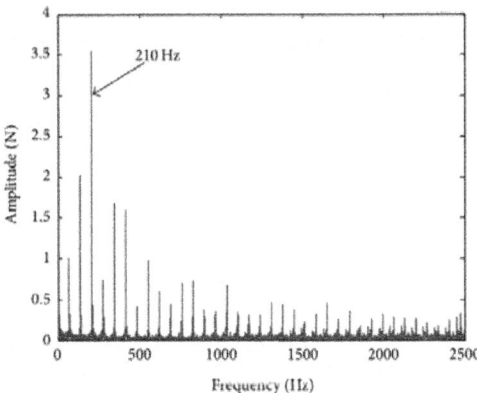

Figure 13: FFT-transform of force signal.

Considering the force and deformation in one wheel revolution, the coupling simulation between grinding wheel and process is performed in a set of harmonic grinding depths with a fixed amplitude of 20 μm and a fixed frequency of 210 Hz, which corresponds to an eigenfrequency of wheel-spindle structure. Figure 14shows that the effectively applied force and the resulting deformation oscillate with the same frequencies and they are inversely influenced by each other. The frequency corresponding to the eigenfrequency of the machine may lead to a nonstable process. Figure 15 shows that through coupling the effective force signal acquires smaller amplitudes than the initial one. The effective force acting at the grinding wheel is smaller than the value corresponding to the uncoupled solution, and the effective force is only about 47% of the initial force in the first cycle which is most likely due to the deformation of grinding wheel. When a grinding wheel experiences deformations the actual depth of cut generated will be smaller, which subsequently causes a smaller grinding force value.

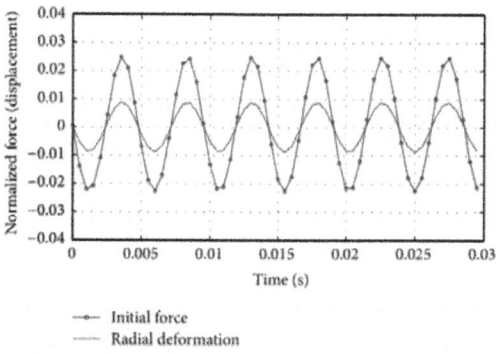

Figure 14: Effective force and displacement.

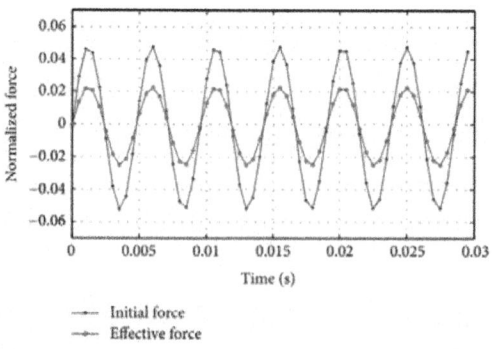

Figure 15: Initial force and effective force.

CONCLUSIONS

In this paper, a method of modeling and simulation for process-machine interaction in grinding of cemented carbide indexable inserts was presented.

A center coordinate adjusting method and a virtual grid method were adopted to model the topographies of grinding wheel. The methods proposed in this work take account of the random nature of abrasive grains and were able to avoid overlapping between abrasive grains in the binder in modeling. With the grinding wheel model at hand, a process model base on KSIM was able to generate forces as input data in process-machine interaction. A wheel-spindle structure was selected and modeled by means of finite element method. The dynamic characteristic of the model was then verified by experimental modal analysis, which proves to match well with the experimental results.

The characteristic equation of the closed-loop dynamic grinding system was derived to account for the inner-relation of process and machine. A coupling simulation with an iteration algorithm was proposed to investigate the process and machine interaction. In the coupling approach, process and machine structure were dealt with in an integrated manner and interacted in synchronized cycles. Dynamic interaction of grinding forces and grinding wheel deformations were then investigated based on the proposed simulation method. It shows that the grinding wheel deformations have an influence on the cutting forces. The coupling method serves as a useful tool to understand well the interaction phenomenon in grinding of cemented carbide indexable inserts.

ACKNOWLEDGMENT

This work was financially supported by the Science and Technology Innovation Project of Shannxi Province under Grant no. 2014KTDZ01-04, and the Science and Technology Major Special Project of China under Grant no. 2011ZX04003-021.

REFERENCES

1.   Z. W. Zhong and V. C. Venkatesh, "Recent developments in grinding of advanced materials," The International Journal of Advanced Manufacturing Technology, vol. 41, no. 5-6, pp. 468–480, 2009.

2.   E. Brinksmeier, Y. Mutlugünes, F. Klocke, J. C. Aurich, P. Shore, and H. Ohmori, "Ultra-precision grinding," CIRP Annals—Manufacturing Technology, vol. 59, no. 2, pp. 652–671, 2010.

3.   C. Brecher, M. Esser, and S. Witt, "Interaction of manufacturing process

and machine tool," CIRP Annals—Manufacturing Technology, vol. 58, no. 2, pp. 588–607, 2009.

4.   J. C. Aurich, D. Biermann, H. Blum et al., "Modelling and simulation of process: machine interaction in grinding," Production Engineering, vol. 3, no. 1, pp. 111–120, 2009.

5.   Y. Lei, J. Lin, D. Han, and Z. He, "An enhanced stochastic resonance method for weak feature extraction from vibration signals in bearing fault detection," Proceedings of the Institution of Mechanical Engineers, Part C: Journal of Mechanical Engineering Science, vol. 228, no. 5, pp. 815–827, 2014.

6.   J. W. Xiang, T. Matsumoto, Y. X. Wang, and Z. Jiang, "A hybrid of interval wavelets and wavelet finite element model for damage detection in structures," Computer Modeling in Engineering & Sciences, vol. 81, no. 3-4, pp. 269–294, 2011. · View at MathSciNet

7.   J. Xiang, T. Matsumoto, J. Long, Y. Wang, and Z. Jiang, "A simple method to detect cracks in beam-like structures," Smart Structures and Systems, vol. 9, no. 4, pp. 335–353, 2012.

8.   H. K. Tönshoff, J. Peters, I. Inasaki, and T. Paul, "Modelling and Simulation of Grinding Processes,"CIRP Annals-Manufacturing Technology, vol. 41, no. 2, pp. 677–688, 1992.

9.   E. Brinksmeier, J. C. Aurich, E. Govekar, et al., "Advances in modeling and simulation of grinding processes," CIRP Annals—Manufacturing Technology, vol. 55, no. 2, pp. 667–696, 2006.

10.   Y. Altintas, C. Brecher, M. Weck, et al., "Virtual machine tool," CIRP Annals—Manufacturing Technology, vol. 54, no. 2, pp. 115–138, 2005.

11.   D. A. Doman, A. Warkentin, and R. Bauer, "Finite element modeling approaches in grinding,"International Journal of Machine Tools and Manufacture, vol. 49, no. 2, pp. 109–116, 2009.

12.   T. A. Nguyen and D. L. Butler, "Simulation of surface grinding process. Part 2. Interaction of the abrasive grain with the workpiece," International Journal of Machine Tools and Manufacture, vol. 45, no. 11, pp. 1329–1336, 2005.

13.   Z. B. Hou and R. Komanduri, "On the mechanics of the grinding process—part I. Stochastic nature of the grinding process," International Journal of Machine Tools and Manufacture, vol. 43, no. 15, pp. 1579–1593, 2003.

14.   C. Brecher and S. Witt, "Simulation of machine process interaction with flexible multi-body simulation," in Proceedings of the 9th CIRP International Workshop on Modeling of Machining Operations, pp. 171–178, Bled, Slovenia, 2006.

15. P. Herzenstiel, R. C. Y. Ching, S. Ricker, A. Menzel, P. Steinmann, and J. C. Aurich, "Interaction of process and machine during high-performance grinding: towards a comprehensive simulation concept,"International Journal of Manufacturing Technology and Management, vol. 12, no. 1-3, pp. 155–170, 2007.

16. K. Weinert, H. Blum, T. Jansen, and A. Rademacher, "Simulation based optimization of the NC-shape grinding process with toroid grinding wheels," Production Engineering, vol. 1, no. 3, pp. 245–252, 2007.

17. J. C. Aurich and B. Kirsch, "Kinematic simulation of high-performance grinding for analysis of chip parameters of single grains," CIRP Journal of Manufacturing Science and Technology, vol. 5, no. 3, pp. 164–174, 2012.

18. G. Warnecke and U. Zitt, "Kinematic simulation for analyzing and predicting high-performance grinding processes," CIRP Annals—Manufacturing Technology, vol. 47, no. 1, pp. 265–270, 1998.

19. X. L. Zhang, B. Yao, and W. Feng, "Modeling of virtual grinding wheel based on random distribution of multi abrasive grains and prediction of grinding force," Acta Aeronautica et Astronautica Sinica, 2014.·

20. J. C. Aurich, A. Bouabid, P. Steinmann, et al., "High-performance surface grinding," in Process Machine Interactions, pp. 81–100, Springer, Berlin, Germany, 2013.

Chapter 8

MODELING OF THE FEED-MOTOR TRANSIENT CURRENT IN END MILLING BY USING VARYING-COEFFICIENT MODEL

Mi Xiao[1] Long Wen[1] Xi Li[2] and Liang Gao[1]

[1]State Key Laboratory of Digital Manufacturing Equipment and Technology, Huazhong University of Science and Technology, Wuhan 430074, China

[2]School of Mechanical Science and Engineering, Huazhong University of Science and Technology, Wuhan 430074, China

ABSTRACT

In order to ensure the stability of the machining process, it is vital to control the machining condition during the milling process. While the feed-motor current is related to many physical variables, such as the cutting force and tool wear, we can indicate it as the key variables to monitoring the conditions of the milling process. A predictive model of the feed-motor current amplitude is established in this paper. The change regulation of the transient current amplitude during the milling process is investigated, and the effect of the spindle speed on the transient current amplitude is studied as well. Since the transient current amplitude is time-varying, the predictive model is a typical panel data type. In this case, the varying-coefficient model (VCM), a potential soft computing method, is applied to solve this predictive model. Then several experiments are conducted to evaluate the performance of VCM method. Results show that the predicted values match the experimental value well, and the correctness of the predictive model for transient current amplitude is also validated.

INTRODUCTION

High-speed machining (HSM) has shown a lot of advantages in the manufacturing industries with the development of the computer numerical control (CNC) systems. It provides a way to achieve a higher quality of the work-piece and can increase the productivity of the machining process. However, due to the complexity of the machining process, the control of HSM to ensure the stability during the machining process is a challenging work.

Many researchers focus on the modelling of the machining process and the physical variables, such as cutting force [1], chatter [2], and tool wear [3], in order to monitor the machining process. In recent years, soft computing methods have been applied to the modeling and prediction in the machining field [4], and their potential is validated. Surface roughness, cutting force, tool life, and the material remove rate are the most common machining physical variables. Moghri et al. [5] applied the artificial neural network (ANN) to predict the surface roughness in milling of ployamide-6 nanocomposites. Ren et al. [6] presented a subtractive clustering-based fuzzy approach method for modelling the cutting force in micromilling; the obtained results proved that the proposed solution can model the cutting force in spite of uncertainties in the micromilling process. Gokulachandran and Mohandas [7] compared neurofuzzy logic technique and support vector regression technique for the assessment of remaining useful life of cutting tools and obtained good results.

As the machining parameters are easier to control compared with other input, the effect of the machining parameters to the physical variables is critical to monitor the machining process, and the application of soft computing method to model the relationship between the machining parameters and machining process is increasing. But most machining physical variables are nonlinear and random, making it difficult to model the machining process. Cakir et al. [8] tried to establish the quantitative relationship between the process parameters and the performance of electrical discharge machining (EDM) process using adaptive neurofuzzy inference system (ANFIS), genetic expression programming (GEP), and ANN. The result shows that all approaches are successful in the prediction of EDM performance. Zohourkari et al. [9] investigated the effects of the machining parameters on the material removal rate in the abrasive water-jet turning process using analysis of variance (ANOVA), and the result is good. Sharkawy [10] presented the study of modeling the surface roughness in end milling using radial basis function neural networks (RBFNs), ANFIS, and genetically evolved fuzzy inference systems (G-FISs), and the prediction accuracy is as high as 97.05%.

Modeling and controlling the machining process using soft computing have achieved a great success. However, even though several machining physical properties have been studied, the research on the field of the feed-motor current remains limited. The feed-motor current is one of the most vital machining variables in the milling process. Many other machining physical variables, such as cutting force and tool wear, are related to the feed-motor current. Aggarwal et al. [11] identified the cutting torque and tangential cutting force coefficient from the spindle motor current, Jeong and Cho [12] estimated the cutting force from the rotating and stationary feed-motor current, and Rizal et al. [3] detected the tool breakage in CNC high-speed milling based on the feed-motor current. Kim and Jeon [13] show that feed-motor is better for the control of the CNC milling process than cutting forces. Thus, the research on the feed-motor current is very essential in order to understand and control the machining process. Xu and Jie [14] studied the current empirical equation between the effective values of feed-motor current and the machining parameters. Even though this empirical equation fits the experiments well, the law of the feed-motor current lacks.

In this research, the feed-motor current is studied using soft computing method. A quantitative prediction model of the feed-motor transient current amplitude is established in the end milling process. The transient current is time-varying during the milling process, making it different from the traditional current empirical equation. The transient current predictive model is extended from this empirical equation. The change regulation of the feed-motor transient current is investigated in the predictive model, as well as the effect of the spindle speed to the feed-motor current. As the feed-motor transient current is time-varying, the transient current predictive model is a typical panel data in this study. Varying-coefficient model (VCM) is applied to solve the transient current predictive model. Several experiments are conducted to evaluate the correctness of the predictive model and the performance of the VCM methods; the result shows that the predictive results match the experimental value very well.

The rest of this paper is organized as follows. In Section 2, the predictive model of the transient current is established. The varying-coefficient model is shown in Section 3. The experiments and results are presented in Section 4. The conclusion is drawn in Section 5.

THE PREDICTIVE MODEL OF THE TRANSIENT CURRENT

The experiments are carried out on a machining center (OTM-650), the work-piece material is the 45 steel and the tool material is high-speed steel, and the number of the tool teeth is three. During the experiments, the spindle speed levels are 500, 600, 700, 800, 1000, and 1200 rpm. The feed speed is 200 mm/min and the axial depth of cut is 0.2 mm in all experiments. There are 10 repeated experiments under the different spindle speed level. As shown in Figure 1, two phases of the feed-motor current, namely, the U phase I_u and the V phase I_v, are sampled during the milling process, and the sampling frequency is 512 Hz.

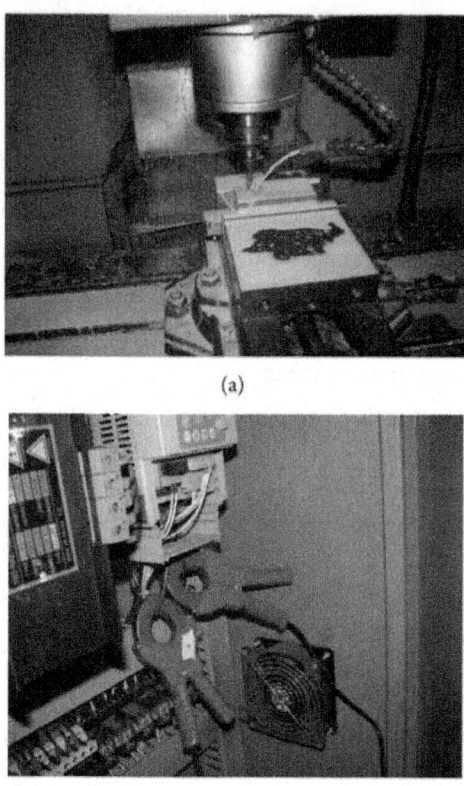

(a)

(b)

Figure 1: The experiment setup.

DATA ACQUISITION

During the milling process, the U phase I_u and the V phase I_v of the feed-motor transient current are acquired and they are enough to calculate the amplitude I of the three-phase feed-motor current. The relationship between them is

$$I_u = I \cos(\omega_1 t),$$

$$I_v = I \cos\left(\omega_1 t + \frac{2\pi}{3}\right),$$

$$I = \sqrt{\frac{4}{3}\left(I_u^2 + I_v^2 + I_u I_v\right)},$$

(1)

where ω_1 stands for the motor frequency, which is related to the spindle speed S, and $\omega_1 = S/22.5$.

Figure 2 shows the samples of the feed-motor transient current amplitude and indicates that the current amplitude is periodically changed, and the cycle is stable. Some feature values can be extracted in order to reduce the data volume of the transient current without losing its properties.

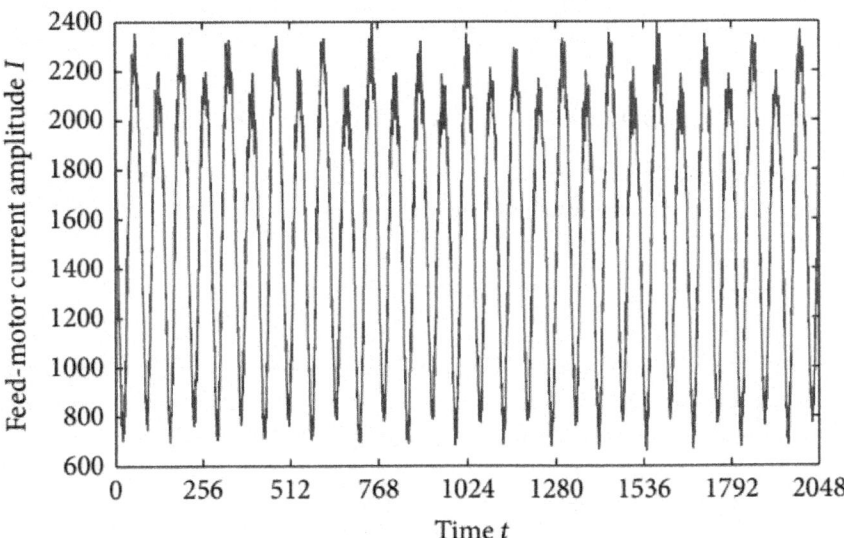

Figure 2: The feed-motor transient current amplitude samples with spindle speed 500 rpm.

Feature Values of the Transient Current

Feature Values

Two feature values are extracted to represent the transient current amplitude, and they are the cycle and the starting point of the transient current. These two feature values are shown as in Figure 3. The cycle is the most important feature of the transient current amplitude. However, the start point is also vital and essential in order to determine the starting phase position of the transient current amplitude. With the help of feature values, it provides a more convenient way for the comparison process, and the analytical results made in one cycle are suitable for other cycles too.

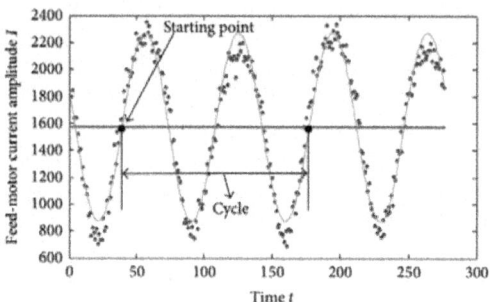

Figure 3: The feature values of the transient current amplitude.

The calculation of the two feature values is based on the Fourier series. The Fourier series is shown as follows:

$$F(x) = a_0 + \sum_{i=1}^{n} \left(a_i \cos(wx) + b_i \sin(wx) \right).$$

(2)

The cycle $T = 2\pi/w$, and the start phase is set to $-\pi/2$. Here, frequency w is a feature of transient current amplitude, and it has no association with the motor frequency w_1. As shown in Figure 3, the 2nd Fourier series is preferred.

Normalization Process

Even though the starting point is used to simplify the comparison of the transient current amplitude, the normalization process is needed. The milling process is disturbed by many factors, and there is a subtle difference between the results of feature values of 10 repeated experiments; even the machining parameters are the same for the repeated experiments.

Table 1 presents the frequency results of 10 repeated experiments using Fourier series. The result shows that they are slightly different from each other. In order to unify the 10 repeated experiments, all cycles are normalized to 1 to eliminate the disturbance.

Table 1: The cycle of the transient current amplitude with spindle speed 500 rpm

Number	1	2	3	4	5	6	7	8	9	10
w	23.25	23.25	23.25	23.24	23.26	23.25	23.25	23.26	23.26	23.24

The Validation Process

The validating process here is to check the effectiveness of the feature values and the normalization process. There are two testing parts in this section. The first one is used for testing the correctness of the feature values, and the second one is to test the validity of the normalization process.

The first testing part is the shift operator. In this operator, the samples in the later four cycles are shifted to the first cycle. The shifted result is shown in Figure 4(a); the black, green, red, pink, and blue samples came from different cycles. The samples in these five cycles are mixed very well, indicating the correctness of the feature values. It also can be inferred that the analytical result in one cycle is suitable for other cycles too.

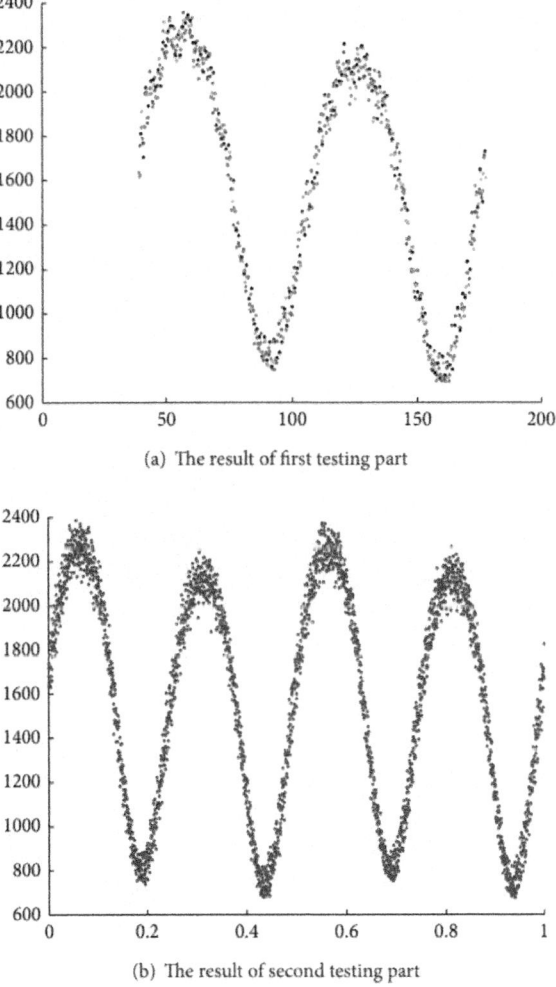

(a) The result of first testing part

(b) The result of second testing part

Figure 4: The validation process of the transient current amplitude with spindle speed 500 rpm.

The second testing part is conducted to test the normalization process. In this round, the cycle is set to 0.5 and the samples are selected on the first two cycles in all 10 repeated experiments with the spindle speed of 500 rpm. As shown in Figure 4(b), all samples are mixed together, and the mixed result is a smooth heavy line. The result indicates that the normalization process matches the experimental data well, meaning the analytical results of the mixed samples are suitable for all 10 repeated experiments.

Transient Current Predictive Model

The feature values show the change regulation of the transient current amplitude during the milling process. In this section, the relationship of the transient current amplitude and the spindle speed is considered, and the predictive model is established.

The traditional relationship between the feed-motor current and the machining parameters is an empirical equation, shown in (3). I_e is the effective value of the feed-motor current, S, F, D are the spindle speed, feed, and depth of cut, and K_0, b_1, b_2, b_3 are the coefficient parameters. Consider

$$I_e = K_0 S^{b_1} f^{b_2} D^{b_3}.$$

$$(3)$$

The transient current amplitude is different from the effective value of current; the former one is changing with time during the machining process. As presented in Section 2.2, the transient current amplitude is changing periodically during the milling process, and the cycle is stable. Moreover, the transient current amplitude is also related to the machining parameters, namely, the spindle speed. In order to consider the dynamic properties of the current amplitude, transient current predictive model extended the empirical equation as follows:

$$I(t) = K_0(t) S^{b_1(t)}.$$

$$(4)$$

The transient current amplitude (t) and coefficient parameters K_0 and b_1 are extended to the function of time, in order to show that the transient current amplitude is time-varying during the stable milling process. The spindle speed affects the transient current amplitude as well in this model. For simplification, the predictive model can be transferred to a linear equation:

$$\ln I(t) = \ln K_0(t) + b_1(t) \ln S.$$

$$(5)$$

The simplified predictive model is a typical panel data type. The current amplitude is affected by the time and the spindle speed, which is the cross-sectional data type. Varying-coefficient model (VCM) is very suitable to handle with panel data, so the VCM method is applied to solve this predictive model.

VARYING-COEFFICIENT MODEL

The traditional statistical model uses sample data (X_i, Y_i) to estimate a regression function in which the response variable Y is represented as a function of the predictor variable X, where Y is one-dimensional and X is a p-dimensional vector $(X = (X_1, X_2 \cdots X_p))$. Usually a linear relationship between Y and X is assumed:

$$Y_j = \beta_j X_j + \varepsilon_j.$$

$$(6)$$

The ε represents the random error and it satisfies the independent and identically distributed condition.

In recent years, several nonparametric regression methods are proposed in order to get a more powerful regression model, such as Nadaraya-Watson method [15, 16] and local linear method [17]. In traditional statistical models, the regression function is usually predefined and we just need to find the optimum coefficients of the regression model. However, the nonparametric regression focuses on finding the best model which fits the experimental data the most,. So the nonparametric regression is a sort of data-driven modeling method and it is more potential to model the nonlinear system [18, 19].

Even though the nonparametric regression method is potential, its application in the machining field is little. Valentinčič and Junkar [20] applied the nonparametric method to the electrical discharge machining process (EDM) and obtained a valuable result. Munoz-Sánchez et al. [21] studied the inverse identification of material parameters using hybrid FEM/LPR method and FEM/ANN. Local polynomial regression (LPR) is also a kind of nonparametric regression. The result shows that LPR method is better than ANN method, indicating the performance of the nonparametric regression method.

The varying-coefficient model (VCM) [22] is an extension of nonparametric regression. It is based on the linear regression and assumes that the coefficient term is the function of another predictor variable U to enhance its flexibility, and the form is

$$Y = \beta_1 (U) X_1 + \beta_2 (U) X_2 + \cdots + \beta_p (U) X_p + \varepsilon.$$

$$(7)$$

In the VCM method, the coefficient (U) is the function of the predictor variable U and is also the coefficient of another predictor variable X_j. The VCM method concentrates more on U than on X. Usually, assume $X_1 \equiv 1$ to add the intercept term.

The estimation of the VCM method is shown in Fan and Zhang [23]. The main task is to use the sample $(Yi; X_{i1}, X_{i2},\ldots, U_i)$ $(i = 1, 2, \ldots, n)$ to minimize the following function

$$\sum_{i=1}^{n} \left\{ Y_i - \sum_{j=1}^{p} \left(\beta_j(u_0) + \beta_j'(u_0)(U_i - u_0) \right) X_{ij} \right\}^2$$

$$\times K_h(U_i - u_0) = \min, \tag{8}$$

where $K_h(t) = K(t/h)/h$ is the kernel function and h is the bandwidth. Gaussian function is a commonly used kernel function, which has the following expression $(t) = \exp(-0.5t^2)/\sqrt{2\pi}$.

Let

$$X(u_0)$$

$$= \begin{bmatrix} X_{11} & \cdots & X_{1p} & X_{11}(U_1 - u_0) & \cdots & X_{1p}(U_1 - u_0) \\ X_{21} & \cdots & X_{2p} & X_{21}(U_2 - u_0) & \cdots & X_{2p}(U_2 - u_0) \\ \vdots & & \vdots & \vdots & & \vdots \\ X_{n1} & \cdots & X_{np} & X_{n1}(U_n - u_0) & \cdots & X_{np}(U_n - u_0) \end{bmatrix},$$

$$W(u_0)$$

$$= \text{diag}\left(K_h(U_1 - u_0), K_h(U_2 - u_0), \ldots, K_h(U_n - u_0) \right),$$

$$a(u_0) = \left(\beta_1(u_0), \beta_2(u_0), \ldots, \beta_p(u_0), \beta_1'(u_0), \right.$$

$$\left. \beta_2'(u_0), \ldots, \beta_p'(u_0) \right)^T$$

$$Y = (Y_1, Y_2, \ldots, Y_n)^T. \tag{9}$$

Assuming that e_j denotes a 2p-dimensional row vector, in which the j-th element is one and the other is zero, the estimation result of $\hat{\beta}_j(\cdot)$ is

$$\hat{\beta}_j(u_0) = e_j a(u_0)$$

$$= e_j \left(X^T(u_0) W(u_0) X(u_0) \right)^{-1}$$

$$\times X^T(u_0) W(u_0) Y. \tag{10}$$

The bandwidth selection is vital to the nonparametric regression, as shown by Härdle and Mammen [24]. Cross-validation [25] is a popular method to evaluate the risk of the model, and there are several cross-validation methods, such as threefold cross-validation, tenfold cross-validation, and leave-one-out cross-validation. However, as shown in [26], the tenfold cross-validation has become the standard method in the practical terms. In this research, the 10-fold cross-validation is preferred. From (7), if the predictor variable X is the machining parameters, the predictor variable U is the time t, and the response variable is the transient current amplitude, the VCM method would have the same formulation with the transient current predictive model, meaning that the VCM method is suitable to solve this model.

EXPERIMENT AND RESULT

Several experiments are conducted under different spindle speed, and the feed-motor current data has been collected. The details of the experiments are described in Section 2.1. In this section, only the first cycle of the transient current amplitude is analyzed.

Model Analysis

The transient current predictive model has the same formulation as the VCM method, but the VCM method is based on the linear regression. It is important to test the linearity of predictor variable ln (t) and the response variable ln S in the predictive model. The samples used in the linearity testing are lain on the same phase in the cycle under different spindle speed level, and the phase e t = 0.02i, i = 0, 1, . . . , 50, as shown in Figure 5. Since the amplitude cycles of all spindle speed level are normalized to 1, the samples can reflect the law of transient current under different spindle speed, and the linearity testing results of ln (t) and ln S are shown in Figure 6.

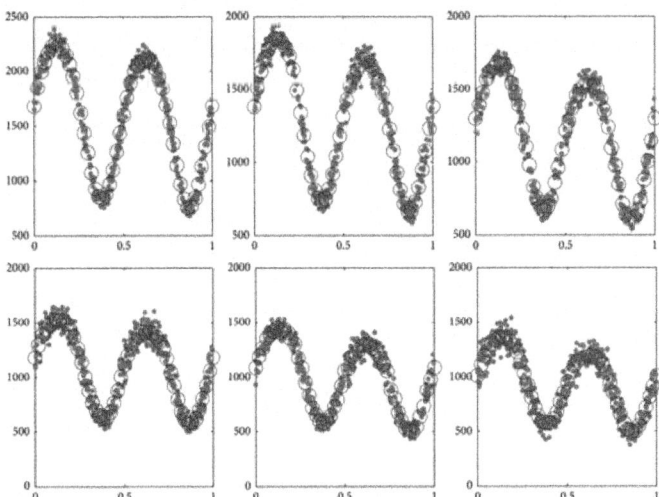

Figure 5: The samples used for the linearity testing of the transient current amplitude.

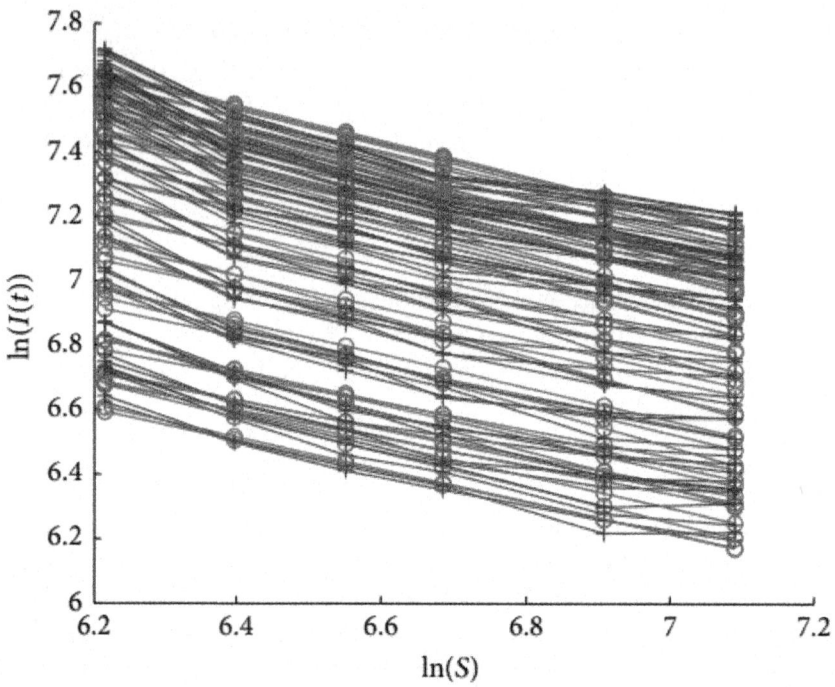

Figure 6: The linearity testing results under each sample for the transient current amplitude.

Table 2 shows the multiple correlation coefficients (R^2) of the linearity result; it can be clearly seen that the mean of R^2 is as high as 0.9374, which indicates that the linearity between ln (t) and ln S is very good. At the same time, the predictive model of the transient current is also validated.

Table 2: The linearity testing result of the transient current amplitude

	Min	Mean	Max
R^2	0.8886	0.9374	0.9793

Results Analysis and Discussion

This section presents the predictive result of VCM method. 600 samples of transient current are randomly selected from the six-level spindle speed. The bandwidth of VCM method is determined using tenfold cross-validation. The results are shown in Figures 7 and 8.

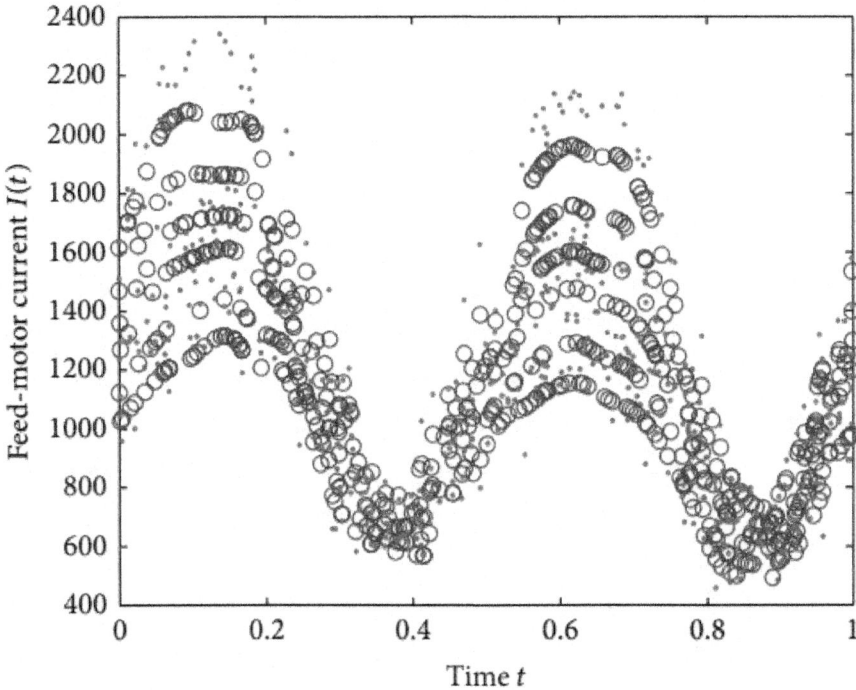

Figure 7: The predictive result of VCM method.

Figure 7 presents the regression result of the total samples, the regression results match the experiment data well, R^2 of the total sample is as high as 0.9806, and the mean relative errors (MRE) are 5.99%, less than 6%. Figure 8 shows the predicted results under different spindle speed and the results also agree with the experimental result well, which indicates that the VCM method achieves good results at all spindle speed levels.

Table 3 presents the comparison between the VCM method and the Fourier series fitting. The Fourier series fitting is used to fit the transient current amplitude on each level of spindle speed. The testing result shows that all the R^2 of VCM are much higher than Fourier series fitting, but these MRE are close to each other. The result indicates that VCM has a better performance to fit the transient current amplitude than the Fourier series fitting because the data samples of VCM came from all spindle speed level; the fault tolerance of VCM is better.

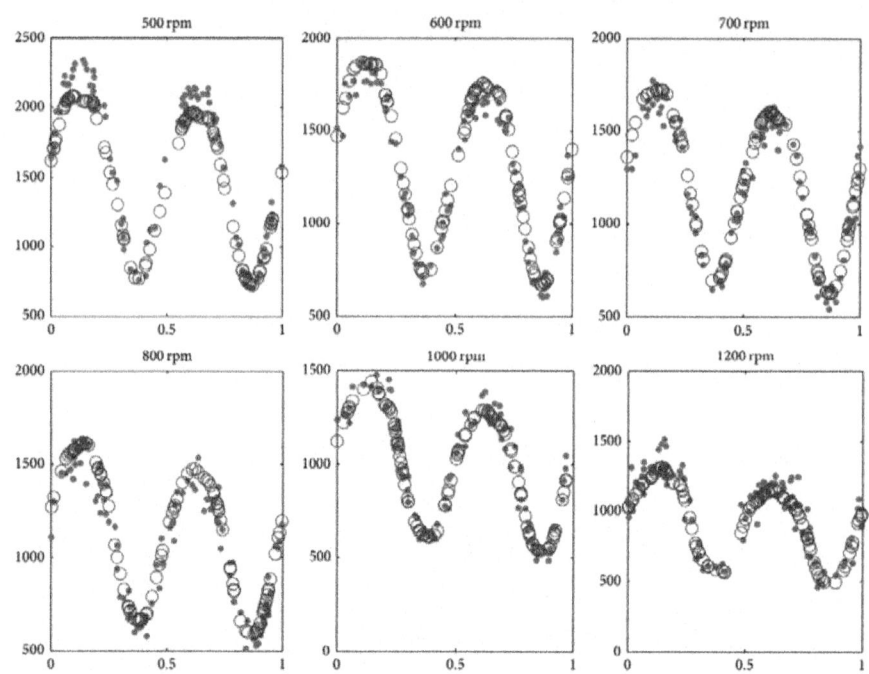

Figure 8: The predictive results of the VCM method under each spindle speed.

Table 3: The result of VCM method

Spindle speed		Total	500	600	700	800	1000	1200
VCM	R^2	0.9806	0.9910	0.9900	0.9876	0.9763	0.9768	0.9647
	MRE	5.99%	6.19%	4.64%	5.91%	6.63%	5.41%	7.14%
Fourier series	R^2	—	0.9690	0.9641	0.9548	0.9348	0.9378	0.8924
	MRE	—	5.36%	5.21%	5.80%	6.76%	6.71%	8.56%

Figure 9 presents the result of the coefficient parameters $\beta_0(t)$, $\beta_1(t)$ of VCM method. $\beta_0(t)$, $\beta_1(t)$ are also the coefficient parameters $\ln K0(t)$, $b1(t)$ of the transient current predictive model. The curves of both coefficients have a big change at the point of 0.4 and 0.9, meaning that the transient current amplitudes are changing over time.

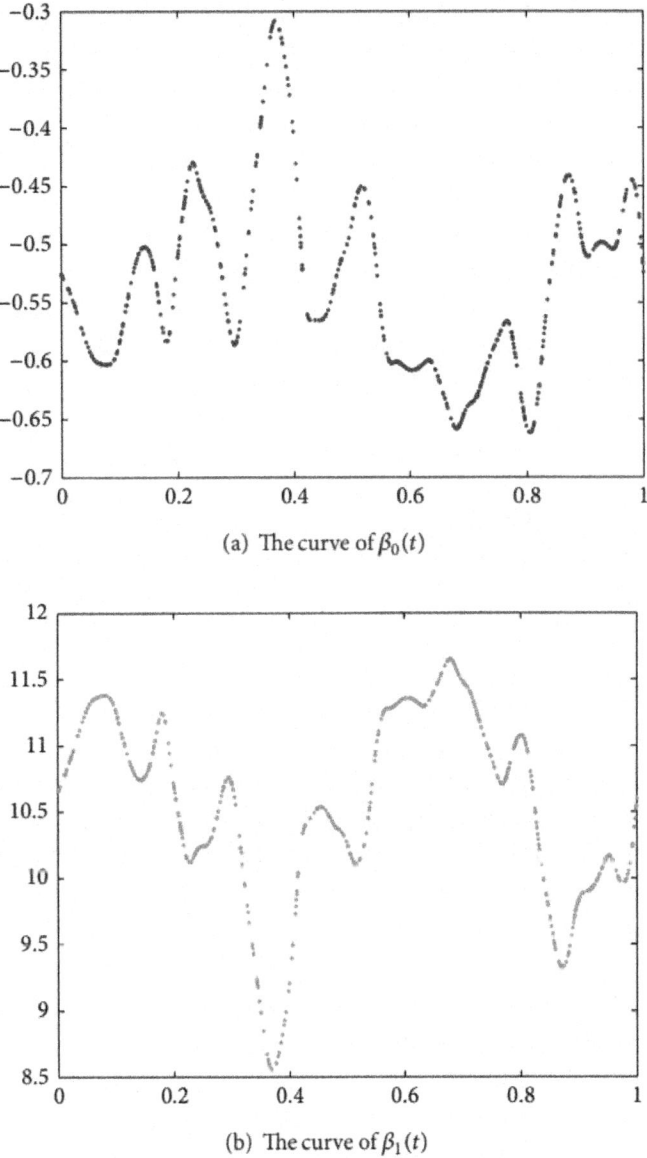

(a) The curve of $\beta_0(t)$

(b) The curve of $\beta_1(t)$

Figure 9: The result of the coefficient parameter of VCM.

From the predictive results and linearity testing, the correctness of the transient current predictive model is validated and the performance of the VCM methods is demonstrated. The predicted results of the transient current amplitude are good and fit the experiment data very well.

CONCLUSION AND FUTURE WORK

The feed-motor current has a wide influence on other cutting variables, such as the cutting force and the tool wear, but there are few studies on the feed-motor current. This paper focuses on the transient current of the stable milling process and established a predictive model for the transient current amplitude. In this model, the change regulation of the transient current amplitude and the effect of the spindle speed to the transient current amplitude are considered. The VCM method is applied to solve this model, because the predictive model is a typical panel data type. Several experiments are conducted to verify the correctness of the predictive model; the results show that VCM method has achieved a good result and the predictive values match the experimental value well.

The main contribution of this paper is the model on the transient current amplitude, but we only focus on the stable milling process. Future works would be done to investigate the relationship between machining parameters and other physical variables in the milling process. As the transient current amplitude is easily interfered by environment factors, the analysis of this disturbance based on VCM method is also important to find more properties of the transient current.

ACKNOWLEDGMENTS

This research work is supported by the National Basic Research Program of China (973 Program) under Grant no. 2011CB706804 and the Natural Science Foundation of China (NSFC) under Grants nos. 51435009 and 51305147.

REFERENCES

1.   D. Cica, B. Sredanovic, G. Lakic-Globocki, and D. Kramar, "Modeling of the cutting forces in turning process using various methods of cooling and lubricating: an artificial intelligence approach," Advances in Mechanical Engineering, vol. 2013, Article ID 798597, 8 pages, 2013.

2.   G. Quintana and J. Ciurana, "Chatter in machining processes: a review," International Journal of Machine Tools and Manufacture, vol. 51, no. 5, pp. 363–376, 2011.

3.   M. Rizal, J. A. Ghani, M. Z. Nuawi, and C. H. C. Haron, "Online tool wear prediction system in the turning process using an adaptive neuro-fuzzy inference system," Applied Soft Computing Journal, vol. 13, no. 4, pp. 1960–1968, 2013.

4.   M. Chandrasekaran, M. Muralidhar, C. M. Krishna, and U. S. Dixit, "Application of soft computing techniques in machining performance

prediction and optimization: a literature review," International Journal of Advanced Manufacturing Technology, vol. 46, no. 5–8, pp. 445–464, 2010.

5. M. Moghri, M. Madic, M. Omidi, and M. Farahnakian, "Surface roughness optimization of polyamide-6/nanoclay nanocomposites using artificial neural network: genetic algorithm approach," The Scientific World Journal, vol. 2014, Article ID 485205, 7 pages, 2014.

6. Q. Ren, M. Balazinski, K. Jemielniak, L. Baron, and S. Achiche, "Experimental and fuzzy modelling analysis on dynamic cutting force in micro milling," Soft Computing, vol. 17, no. 9, pp. 1687–1697, 2013.·

7. J. Gokulachandran and K. Mohandas, "Comparative study of two soft computing techniques for the prediction of remaining useful life of cutting tools," Journal of Intelligent Manufacturing, 2013.

8. M. V. Cakir, O. Eyercioglu, K. Gov, M. Sahin, and S. H. Cakir, "Comparison of soft computing techniques for modelling of the EDM performance parameters," Advances in Mechanical Engineering, vol. 2013, Article ID 392531, 15 pages, 2013.

9. I. Zohourkari, M. Zohoor, and M. Annoni, "Investigation of the effects of machining parameters on material removal rate in abrasive waterjet turning," Advances in Mechanical Engineering, vol. 2014, Article ID 624203, 11 pages, 2014.

10. A. B. Sharkawy, "Prediction of surface roughness in end milling process using intelligent systems: a comparative study," Applied Computational Intelligence and Soft Computing, vol. 2011, Article ID 183764, 18 pages, 2011.

11. S. Aggarwal, N. Nešić, and P. Xirouchakis, "Cutting torque and tangential cutting force coefficient identification from spindle motor current," International Journal of Advanced Manufacturing Technology, vol. 65, no. 1–4, pp. 81–95, 2013.

12. Y. H. Jeong and D. W. Cho, "Estimating cutting force from rotating and stationary feed motor currents on a milling machine," International Journal of Machine Tools and Manufacture, vol. 42, no. 14, pp. 1559–1566, 2002.

13. D. Kim and D. Jeon, "Fuzzy-logic control of cutting forces in CNC milling processes using motor currents as indirect force sensors," Precision Engineering, vol. 35, no. 1, pp. 143–152, 2011.

14. M. Xu and C. Jie, "Research on the relationship between CNC machine tool spindle current and cutting parameters," Machinery, vol. 12, pp. 28–30, 2010.

15.   G. Dudek, "Tournament searching method for optimization of the forecasting model based on the Nadaraya-Watson estimator," Artificial Intelligence and Soft Computing, vol. 8468, pp. 339–348, 2014.

16.   H. Long and L. Qian, "Nadaraya-Watson estimator for stochastic processes driven by stable Lévy motions," Electronic Journal of Statistics, vol. 7, pp. 1387–1418, 2013.

17.   M. Avery, Literature Review for Local Polynomial Regression, 2013,http://www4.ncsu.edu/~mravery/AveryReview2.pdf.

18.   A. Donnelly, B. Misstear, and B. Broderick, "Application of nonparametric regression methods to study the relationship between NO_2 concentrations and local wind direction and speed at background sites,"Science of the Total Environment, vol. 409, no. 6, pp. 1134–1144, 2011.

19.   F. Li, T. Yan, and L. Su, "Solution of an integral-differential equation arising in oscillating magnetic fields using local polynomial regression," Advances in Mechanical Engineering, vol. 2014, Article ID 101230, 9 pages, 2014.

20.   J. Valentinčič and M. Junkar, "Detection of the eroding surface in the EDM process based on the current signal in the gap," The International Journal of Advanced Manufacturing Technology, vol. 28, no. 3-4, pp. 294–301, 2006.

21.   A. Munoz-Sánchez, I. M. González-Farias, X. Soldani, and M. H. Miguélez, "Hybrid FE/ANN and LPR approach for the inverse identification of material parameters from cutting tests," International Journal of Advanced Manufacturing Technology, vol. 54, no. 1–4, pp. 21–33, 2011.

22.   B. U. Park, E. Mammen, Y. K. Lee, and E. R. Lee, "Varying coefficient regression models: a review and new developments," International Statistical Review, 2013.

23.   J. Fan and W. Zhang, "Statistical methods with varying coefficient models," Statistics and Its Interface, vol. 1, no. 1, pp. 179–195, 2008. ·

24.   W. Härdle and E. Mammen, "Comparing nonparametric versus parametric regression fits," The Annals of Statistics, vol. 21, no. 4, pp. 1926–1947, 1993.

25.   S. Arlot and A. Celisse, "A survey of cross-validation procedures for model selection," Statistics Surveys, vol. 4, pp. 40–79, 2010.

26.   I. H. Witten, E. Frank, and M. A. Hall, Data Mining: Practical Machine Learning Tools and Techniques, China Machine Press, Beijing, China, 3rd edition, 2012.

Chapter 9

SURFACE LAYER PROPERTIES AFTER SUCCESSIVE EDM OR EDA AND THEN SUPERFICIAL ROTO-PEEN MACHINING

Agnieszka Dmowska Bogdan Nowicki and Anna Podolak-Lejtas

Institute of Manufacturing Technology, Warsaw University of Technology, 02-524 Warsaw, Poland

ABSTRACT

The paper presents the results of the influence of basic electrical discharge machining EDM parameters and electrical discharge alloying EDA parameters on surface layer properties and on selected performance properties of machine parts after such machining but also the influence of superficial cold-work treatment applied after the EDM of EDA on modification of these properties. The investigations included texture of the surface, metallographic microstructure, microhardness distribution, fatigue strength, and resistance to abrasive wear. It was proved that the application of the roto-peen after the EDM and the EDA resulted in lowering roughness height up to 70%, the elevation of surface layer microhardness by 300–700 µHV, and wear resistance uplifting by 300%.

BASICS OF THE EDM, THE EDA, AND THE ROTO-PEEN MACHINING

Electrical discharge machining (EDM) is widely used in manufacturing of accurate parts of complex shapes, built of hard and impossible-to-cut or hard-to-cut materials [1, 2].

Electrical discharge alloying (EDA) is applied for generation of surface layer (SL) with volume share of up to several dozens percent of alloy elements. Such layers are up to 200 µm thick and they can be useful in wear-resistant tooling, erosion-resistant equipment, and so forth [3, 4].

The EDM and EDA processes are defined by spark discharge and the associated physical processes. Such processes take place in the presence of dielectric—in the EDM usually kerosene or water is employed and the spark discharges release energy of several millijoules while the EDA process is carried out in air, neutral gas, oil or kerosene, and the spark discharge energy is much higher and discharge duration is longer. Spark discharge generate much heat on small surface of the anode and the cathode. Power density inside the spark channel has been estimated as 10^{17} W/m^2 and local temperature could be raised up to 20 000 Kelvin degree [1, 3, 5, 6].

The eruption of molten material is critically important for material removal in the EDM process. Spark discharge results in local material loss, the craters and flashes are formed (Figure 1), and they are covered with the thin film of previously molten and then solidified metal [1, 7]. The volume of material removed from a single crater is $v = 1500 \div 3\ 000\ 000\ (\mu m^3)$.

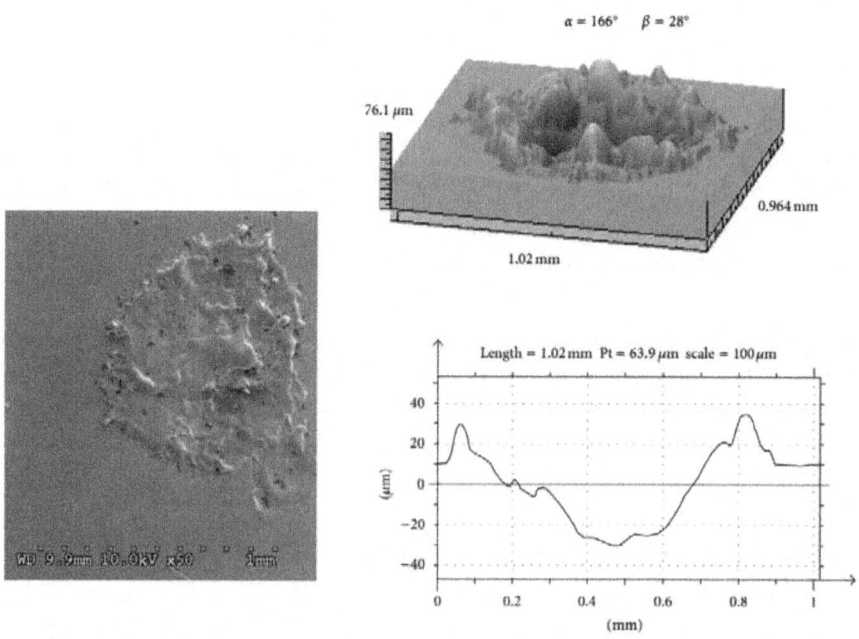

Figure 1: The SEM image of an individual electric discharge on the machined surface, its 3D view, and its profilogram (U = 80 V, I = 48 A, Ti = 400 µs).

Due to repeatedly occurring spark discharges, a surface layer of specific isotropic roughness is being created and this layer shows different metallographic microstructure and microhardness than that of the core. Tensile stress and microcracks are likely be present there in the case of the EDM carried out with medium or high energy of discharges [1, 5, 8, 9].

Two basic processes take place in the EDA and the associated mass transport can be observed as well. The first process consists in crater creation and mass loss on the cathode, while in the other one, mass transport, takes place through the spark channel and formation of highland in the crater (Figure 2), chemical composition of which is enriched by the alloying components [3, 4]. Surface layer (SL) created in the EDA process resembles the surface layer created in the EDM in view of similar phenomena influencing the machining process with one exception—its chemical composition is different. These SL similarities include metallographic structure and microhardness which shows difference when compared to the core, the presence of tensile stress and microcrack occurrence.

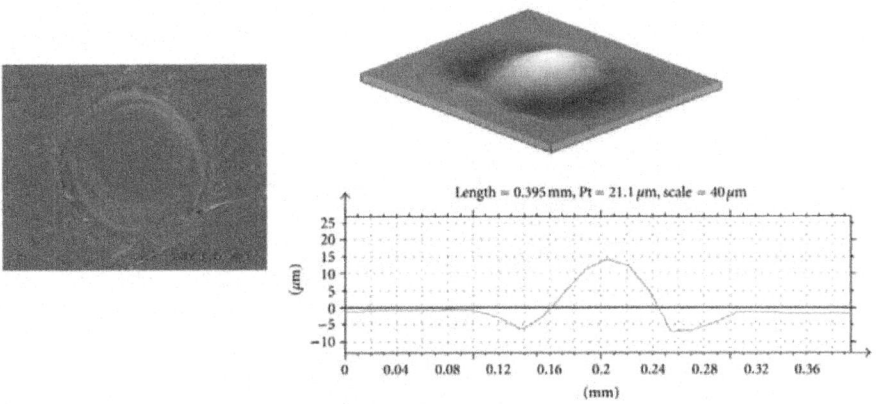

Figure 2: SEM image of individual discharge trace during the EDA on the machined surface, its 3D view and profilogram for U = 160 V, I = 8 A, Ti = 50 μs.

When analyzing process of surface layer constitution, we can isolate three basic types of interaction: thermal phenomena, metallurgic phenomena, and mechanical phenomena.

In the EDM process, thermal and metallurgic phenomena prevail: melting, solidification, microstructural change and their results—change of structure, microhardness, stress, and microcracks.

In the EDA, phenomena resembling those present in the EDM occur but change of chemical composition takes place as well and is connected with spark discharge and with mass transport from the anode to the cathode.

Mechanical processes are typical of superficial cold-work treatment such as roto-peen (R-P) [10–13] and they can result in strain hardening of material (increase in SL hardness), introduction of compressive stress into the SL, elimination or reduction of microcracks which translates into elimination or minimization of adverse phenomena taking place during surface layer formation after the EDM and EDA process [14, 15].

The roto-peen machining process elaborated by the Boeing and 3M company [2, 9] consists in dynamic interaction between little balls of cemented carbide, 1 mm in diameter, fixed to the blades made of glass fiber balls (Figure 3) which rotate with the speed of revolution n = 3 000–8 000 rev/min. Because of the repeated balls against the machined surface, the elastic and plastic deformations follow in surface layer resulting in surface smoothing, superficial strain hardening, introduction of the compressive stress into the surface layer, and increase in fatigue resistance and resistance to abrasive wear. As the size of ball imprint d on the surface is known, one can determine the cold-work depth Z as well as the depth of residual stress existence, which is around 40–80% larger than cold work depth. These numbers should be greater than the depth of the surface layer generated in the EDM and EDA preceding the roto-peen machining.

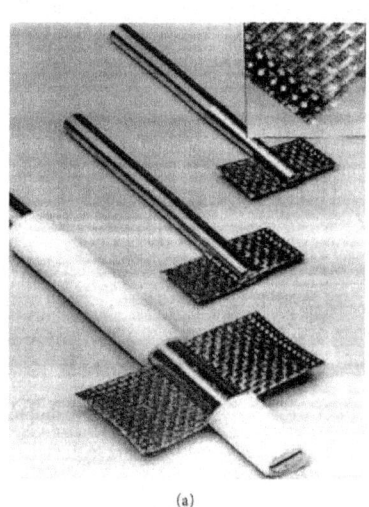

(a)

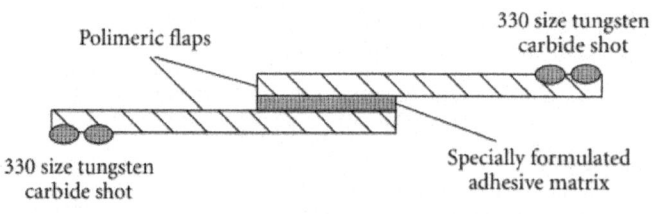

(b)

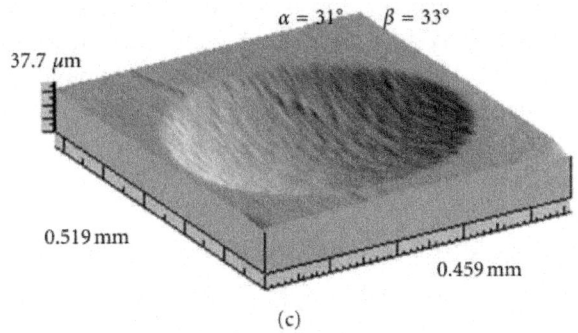

(c)

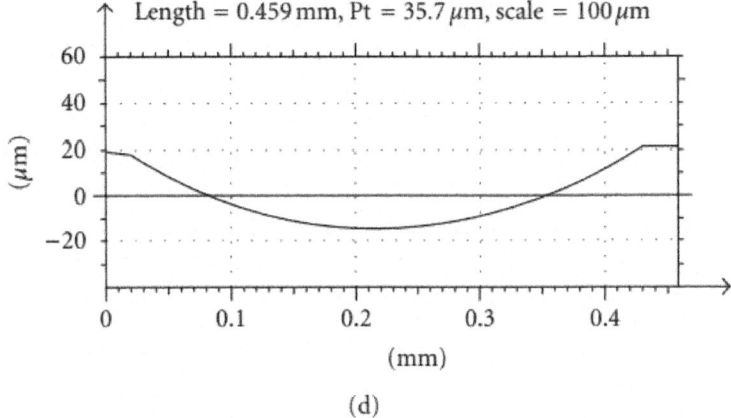

(d)

Figure 3: (a, b) Tools for the roto-peen machining; (c) view of a trace of a single ball hit; (d) its profilogram (Fn = 400 N; 34 HRC).

The impact of a single roto-peen ball interaction on the cold-work depth can be determined (according to Drozd) based on the empirical relationship [10, 12]:

$$Z = 5.5h\frac{D}{d},$$

(1)

where h: depth of the ball imprint, D: ball diameter, and d: diameter of a ball imprint on flat surface.

The research conducted in Warsaw University of Technology, Institute of Manufacturing Technology, was aimed at determining the effectiveness of superficial (roto-peen) cold-work treatment application for elimination of the adverse surface layer features being a consequence of the EDM and EDA.

The experimental investigation included evaluation of surface integrity after the EDM and EDA as well as after combined the EDM + cold work treatment and the EDA + superficial cold work treatment and it concerned the following areas: texture of surface, metallographic microstructure, microhardness distribution, residual stress distribution, fatigue resistance, and resistance to abrasive wear.

RESULTS OF INVESTIGATION FOR SURFACE LAYER PROPERTIES AND SELECTED PERFORMANCE FEATURES AFTER THE EDM AND AFTER COMBINED EDM + ROTO-PEEN TREATMENT

Research has been conducted for typical materials and typical conditions of machining. Samples made of hardened steel NC10 have been subjected to electrical discharge machining using a copper tool electrode mounted on the EDM center ROBOFROM 30 and the following parameters were selected: $U = 160\,V$, $I = 6\,A$, $Ti = 100\,\mu s$, and $U = 160\,V$, $I = 16\,A$, $Ti = 800\,\mu s$, which are typical of finishing machining, then the following parameters: $U = 120\,(V)$, $I = 24\,(A)$, $Ti = 200\,(\mu s)$, which are typical of semifinishing and then $U = 80\,V$, $I = 48\,A$, $Ti = 400\,\mu s$, for the most productive machining type (Table 1).

Table 1: Specification of surface microstereometry (3D) after the EDM and then the roto-peen machining

Parameters EDM			Parameters roto-peen		Microstereometric parameters 3D					
U (V)	I (A)	Ti (μs)	v (rev/min)	t (min)	S_a (μm)	S_t (μm)	S_p (μm)	S_{dq}	S_{sc} ($1/\mu m$)	ST_p (%)
120	24	200	—	—	12.40	87.2	47.5	0.40	0.10	0.80
120	24	200	5000	2.5	4.84	52.9	12.3	0.21	0.02	5.81
120	24	200	5000	5	2.04	27.3	14.2	0.08	0.01	0.00
120	24	200	7000	2.5	6.35	67.6	25.8	0.23	0.03	0.20
120	24	200	7000	5	3.37	41.7	14.6	0.12	0.02	0.30
80	48	400	—	—	13.30	149.0	92.6	0.4	0.03	0.2
80	48	400	5000	2.5	11.30	88.7	22.9	0.31	0.04	4.52
80	48	400	5000	5	6.23	50.4	17.6	0.20	0.03	1.73
80	48	400	7000	2.5	6.19	55	19.8	0.22	0.03	0.40
80	48	400	7000	5	6.51	57.8	17.4	0.16	0.02	3.26
160	16	800	—	—	8.91	78.8	35.4	0.26	0.06	0.60
160	16	800	5000	2.5	5.42	48.1	13.3	0.16	0.01	6.20
160	16	800	5000	5	5.19	47.2	12.8	0.16	0.01	9.55
160	16	800	7000	2.5	5.00	48.1	15.7	0.17	0.02	4.39
160	16	800	7000	5	3.78	53.7	16.4	0.16	0.03	0.60
160	6	100	—	—	4.47	47.4	16.5	0.27	0.01	0.50
160	6	100	5000	2.5	1.31	15.3	15.3	0.05	0.02	0.10
160	6	100	5000	5	1.14	28.5	28.5	0.06	0.02	0.00
160	6	100	5000	7.5	1.36	17	17	0.09	0.02	0.01
160	6	100	7000	2.5	1.66	35.5	35.5	0.08	0.02	0.00

Typical 3D views and profilograms of the EDM-ed and then roto-peened surfaces have been presented in Figures 4 and 5 for various parameters.

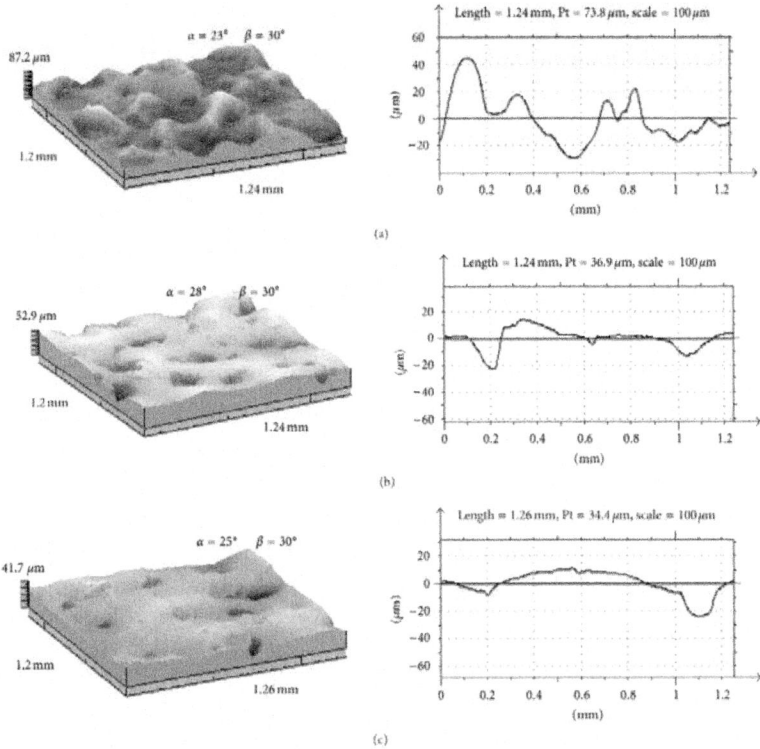

Figure 4: Stereometric image of surface after machining and its profilogram (a) the EDM—U = 120 (V), I = 24 (A), Ti = 200 (μs), and the following roto-peen (b) n = 6500 (rev/min), t = 2.5 (min), (c) n = 6500 (rev/min), t = 5 (min).

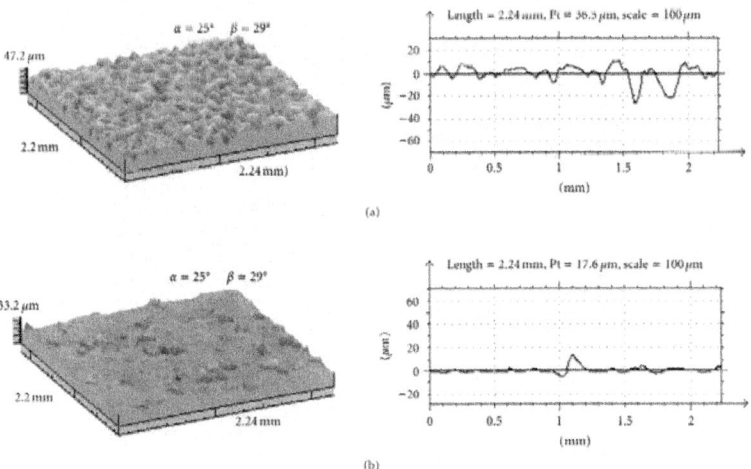

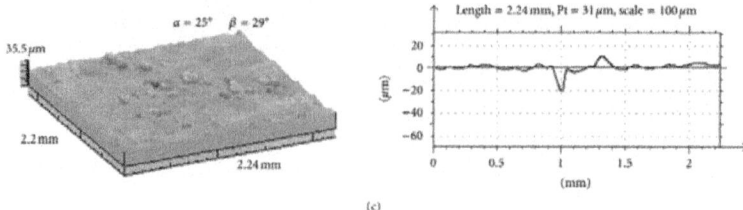

(c)

Figure 5: Stereometric image of surface after machining and its profilogram (a) the EDM– U = 160 (V), I = 6 (A), Ti = 100 (µs), and the following roto-peen (b) n = 6500 (rev/min), t = 2.5 (min), (c) n = 6500 (rev/min), t = 5 (min).

Microgeometry of the roughly EDM-ed and then rotopeened surfaces is typical of wide-spread and smooth summits and of microindents left as remnant of the EDM. Surface texture image in this case does not show much difference for all investigated parameter range. The level of height parameters gets less than one-third the initial value, for example, Ra drops from 21 µm to 6 µm and summit curvatures Ssc from 0.04 to 0.02 (1/µm). Machining time t = 2.5 (min) is needed for a complete smoothing of 10 × 20 surface at n = 7000 (rev/min); t = 5 (min) at n = 5000 (rev/min).

The investigation of surface layer was aimed at assessing the influence of the roto-peen machining on surface layer properties after consecutive EDM and roto-peen machining and particular attention was paid to possible presence of discontinuities, microcracks which can persist in surface layer after the EDM; microhardness distribution were obtained and surface uniformity was assessed. Figure 6 shows images taken as microscopy view of polished sections for specimens machined by the EDM and the roto-peen. Surface layer obtained in the EDM features considerable nonuniformity and microcracks.

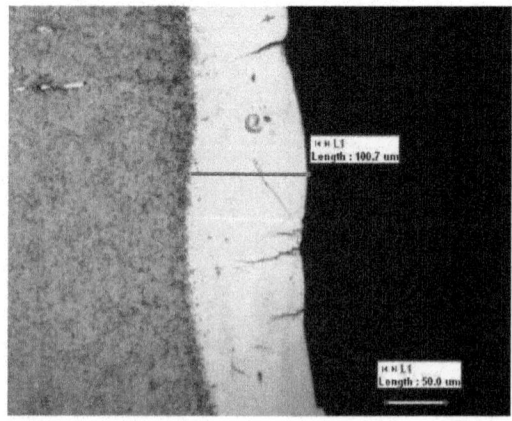

(a)

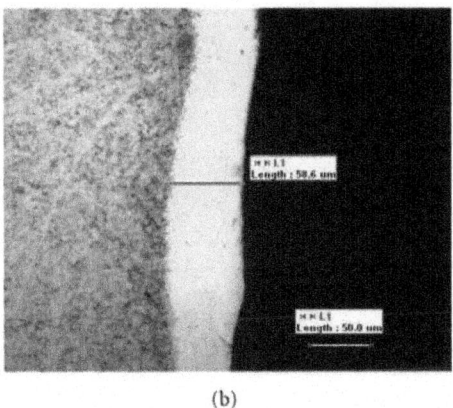

(b)

Figure 6: Surface layer microstructure after the: (a) EDM U = 80 (V), I = 48 (A), Ti = 400 (μs), (b) EDM U = 80 (V), I = 48 (A), Ti = 400 (μs), Tp = 50 (μs) and roto-peen n = 5000 (rev/min), t = 5 (min).

Analysis of the examination for surface layer micro-structure and its thickness after the EDM and the roto-peen machining showed that surface layer is uniform and deprived of any visible micro-cracks or discontinuities in all cases. Surface layer thickness after two consecutive machining runs is much more uniform than after a single-EDM machining.

Microhardness examination has been conducted for the EDM-ed as well as EDM-ed and then roto-peened specimens and the following machining parameters have been applied:

(i) n_1 = 5000 obr/min, t_1 = 2.5 min, t_2 = 5 min,

(ii) n_2 = 7000 obr/min, t_1 = 2.5 min, t_2 = 5 min.

Typical micro-hardness distribution of surface layer after the EDM (Figure 7(a)) and the roto-peen applied after the EDM has been presented on Figure 7(b).

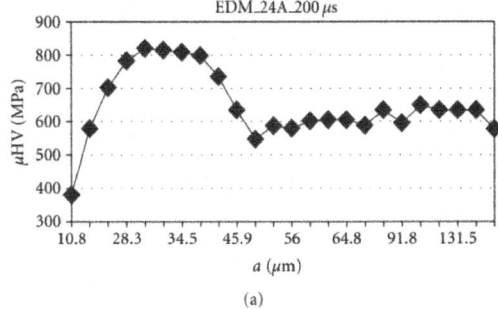

(a)

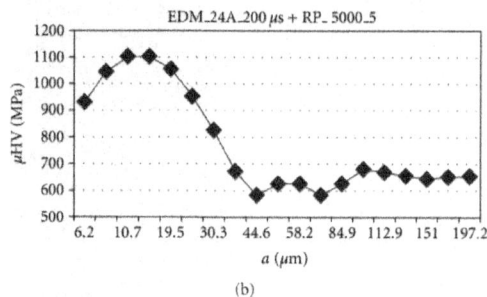

(b)

Figure 7: Microhardness distribution in surface layer for EDM U = 120 (V), I = 24 (A), Ti = 200 (μs), and machined by the EDM and roto-peen n = 5000 (rev/min), t = 5 (min).

Analysis of microhardness distribution of the in-depth surface layer showed that application of the roto-peen machining which follows the EDM results in considerable micro-hardness SL increase from ca. 800 μHV up to 1100 μHV and surface layer improvement as to microgeometry, metallographic structure, and microcrack occurrence.

The residual stress investigations after the EDM and combined EDM + roto-peen machining have shown that after the EDM, the tensile stress taking values up to s = +650 MPa occurs and the application of the roto-peen after the EDM causes shift of this stress into compressive stress taking values up to s = −1200 MPa

The scope of experimental investigation covered two basic performance properties, that is, resistance to abrasive wear and fatigue strength. The experiments of resistance to wear have been performed on Skoda-Savin machine. The assessment of resistance to wear consists in measuring the volume of an indent created on the surface of the examined specimen which is subjected to friction contact with special antispecimen of ring-like shape made of cemented carbide.

The cold-work superficial treatment can function as a factor both increasing and diminishing abrasive wear of the examined specimens depending on how friction is applied [13].

In the case of lubrication-free friction (specimen immersion in potassium dichromate solution), resistance to wear can be lower because adhesive wear is dominant and the surface is likely to be subjected to adhesive blockade with inner energy increase.

In the case of lubrication contact accompanying abrasive wear process, the hardness increase resulting from the cold-work superficial treatment is likely to contribute to higher abrasive wear resistance.

The test has been repeated three times for each specimen and the average measurement value has been calculated. The experimental investigation for examining the influence of the combined EDM + roto-peen machining on resistance to abrasive wear has been carried out for hardened specimen of tool steel NC10 for the following surfaces: ground surfaces, surfaces subjected to the EDM using various parameters (surfaces featuring various layer properties have been investigated), surfaces subjected to the EDM, and then to the roto-peen treatment under various process parameters (time, rotary speed).

The experimental investigation has been carried out under two different ways of experimental procedures: without lubrication, in potassium dichromate solution and for rotational speed n=3000 (rev/min) and then with lubrication in mixed solution of one part of kerosene and three parts of machine oil. The increase of resistance to abrasive wear when compared to the ground specimen is noticeable regardless of the EDM parameters. Resistance to abrasive wear shows a decrease for the EDM-ed specimens when the EDM parameters are elevated. The roto-peen treatment application after the EDM cause increase in wear resistance for friction in potassium dichromate solution.

The diagram presenting resistance to abrasive wear for friction with lubrication has been shown in Figure 8 for the specimen of tool steel, for various machining variants.

Results of investigating resistance to abrasive wear in lubrication friction by the Skoda-Savin method demonstrate considerable increase in wear resistance of the specimens subjected to subsequent EDM and roto-peen machining when compared to the ground or EDM-ed specimens (Figure 8).

Fatigue resistance has been investigated for the flat specimens made of heat-treated tool steel NC10 (Rockwell scale hardness was 30). The Schenk PWQ-flato fatigue testing machine has been used, for two-side bending, with bending stress $\sigma = 300 \div 450$ MPa

Fatigue resistance experiments were limited to general recognition and they were carried out with limited number of variants for specimens machined by the EDM and the EDM+ roto peen for various machining parameters. The specimens after the EDM process carried out for specific parameters have been adopted as a reference condition.

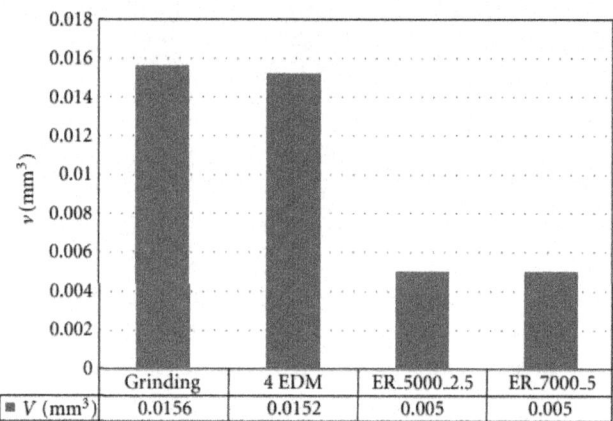

■ V (mm³)	Grinding	4 EDM	ER_5000_2.5	ER_7000_5
	0.0156	0.0152	0.005	0.005

Figure 8: Specification of the resulting resistance to abrasive wear for friction with lubrication of the samples of tool steel NC10 machined by grinding and by the EDM: U = 80 (V), I = 48 (A), Ti = 400 (μs), and EDM + roto-peen: n = 5000 (rev/min), t = 2.5 (min), and n = 7000 (obr/min), t = 5 (min).

The specimens machined by the EDM and then by the roto-peen at parameters n = 5000 (rev/min), t = 2.5 (min) and stress σ = 300 MPa can stand 1 000 000 cycles and for higher stress σ = 350 MPa they can stand 111 000 ÷ 150 000 cycles. Specimens machined by the roto-peen at parameters n = 7000 (rev/min), t = 2.5 (min), and stress σ = 350 MPa or σ = 400 MPa can stand 1 000 000 cycles while for higher stress σ = 450 MPa they can stand 107 000 cycles.

One can conclude that if the resistance of the specimens machined by the EDM for stress 350 MPA amounting to over 100 000 cycles and those machined by the EDM and then the roto-peen showed the resistance 470 000 ÷ 1000 000 cycles, the fatigue resistance after the EDM + roto-peen machining is five to ten times higher for the same stress level. The fatigue strength analysis for the specimens subjected to the EDM and EDM + roto-peen at stress 150 MPa showed that the specimens machined by the EDM show similar resistance for stress σ = 150 MPa that the specimens machined by the EDM and then by the roto-peen for stress σ = 450 MPa

The results obtained show that the roto-peen machining applied after the EDM yields the similar effect as much more intensive shot peening applied after the EDM, described in the available bibliography [12, 14].

THE INVESTIGATION OF THE EDA AND THE ROTO-PEEN INFLUENCE ON SURFACE LAYER PROPERTIES AND SELECTED PERFORMANCE PROPERTIES

The investigation of the EDA parameters on surface layer properties and on the selected performance properties has been carried out for the specimens of hardened constructional steel C45. The experiments have been performed using monolithic shape electrodes made of the following materials: alloy steel 1H18N9, tungsten and tungsten carbide. The assessment included the influence of electrical discharge alloying on surface layer properties. The amperage I within the range of 16–48 A and pulse duration time within 25 do 3200 µs were set as independent variables while constant voltage U = 160 V was used as a fixed parameter. The machining took place in the presence of a typical industrial fluid—cosmetic kerosene. The metallographic structure of surface layer after the EDA is presented in Figure 9.

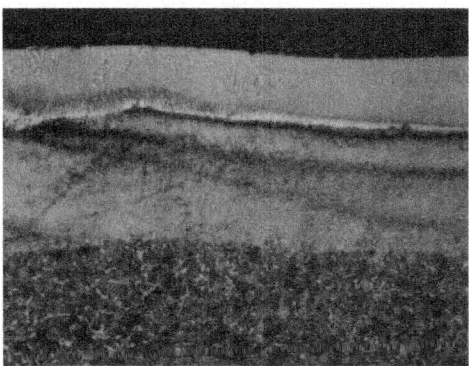

Figure 9: The image of the metallographic structure of surface layer after the EDA: U = 200 (V), I = 32 (A), Ti = 3200 (µs) (x 250).

The selection of the roto-peen parameters insured various strain hardening depths (which are at least equal to the depth of the remelted layer) and various stress level in surface layer, depth of which should be higher than the depth of stress accumulation in surface layer after the EDA and EDM.

The selection of superficial cold work treatment—the roto-peen should be particularly closely watched [14, 16]. Overly intensive machining, both as to the rotational speed and its duration results in destruction of the remelted layer, microcracking and the layer exfoliation while insufficient parameters do not assure proper surface layer depth [4]. After the preliminary investigations, the range of parameters ensuring good machining results has been determined. Rotational tool speed has been set on a level of n = 4100 (rev/min) (maximum allowable speed ensuring that cracks and exfoliation of the alloyed layer would not take place). The times of machining were t1 = 75 (s) and t_2 = 150 (s) (area s = 20 mm^2).

It is assumed that the surfaces to be alloyed had been subjected to the electrodischarge machining on the same machine tool.

Consequently, the investigation does include eight experiments: two EDM variants, two EDA variants, and two roto-peen variants.

The obtained investigation results enable preliminary assessment of the EDA and EDA +roto-peen and determination of its advisability while having known performance requirements of machine parts in mind.

The obtained average values of the remelted surface layer in the EDA process amount from 30 μm for I = 32 (A), Ti = 25 (μs) to 100 μm for I = 48 (A), Ti = 1600 (μs). The analysis of the obtained results indicate intensive growth of the remelted layer with the pulse amperage increase (Figure 10).

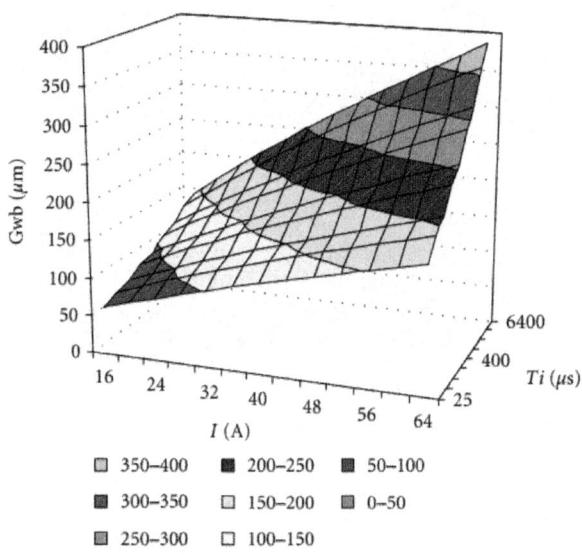

Figure 10: Dependence of the alloyed layer depth on the adopted machining parameters.

The analysis of chemical composition included the alloyed layer but also the heat-affected zone. It has been conducted in vacuum along the direction which is perpendicular to the specimen surface and taken down the surface layer depth. The analysis show that the alloying process employing the electrode 1H18N9 steel gave the remelted layer depth of about 100 μm with high content of alloying elements where the weight shares of the alloyed layers are following: chromium Cr: 6.7%, nickel Ni: 2.8% at the following alloying parameters: pulse amperage: 48 A, pulse duration: 1600 μs (Table 2; Figure 11). For the electrodes with different materials like tungsten and tungsten carbide, saturation of surface layer with the elements coming form the tool electrode is also visible (Table 3). For the alloying process carried out with tungsten electrode at the following parameters: pulse amperage 64 A, pulse time 400 μs, the weight share in surface layer is 6% for tungsten on the average and it is higher amounting respectively up to 33% tungsten for the experiments conducted at elevated pulse amperage 48 A and pulse duration: 1600 μs. Maximum volume share of anode material in the superficial layer of the alloyed part amounted form 20% to 30%.

Table 2: Specification of the chemical composition investigation results for the surface layer after the alloy steel 1H18N9 electrode alloying; pulse current I = 32 (A), pulse Ti = 25 (μs), electrode material: 1H18N9

18L	Si-K	Cr-K	Mn-K	Ni-K
pt1	0.39	5.91	1.09	2.64
pt2	0.35	5.47	1.38	2.60
pt3	0.37	3.59	1.14	2.02
pt4	0.33	4.85	1.33	1.40
pt5	0.38	0.50	0.92	0.00
pt6	0.25	0.38	0.51	0.16
pt7	0.28	0.29	0.75	0.15
pt8	0.42	0.09	1.20	0.44
pt9	0.36	0.24	0.63	0.53
pt10	0.35	0.19	1.05	0.04

Table 3: Specification of the chemical composition investigation results for the surface layer after the tungsten electrode alloying. Pulse current I = 32 (A), pulse Ti = 25 (μs), electrode material: tungsten

26L	Si-K	Mn-K	W-L
pt1	1.52	0.46	33.81
pt2	1.02	0.69	28.38
pt3	1.19	0.49	17.10
pt4	1.08	0.10	31.06
pt5	0.11	0.44	1.37
pt6	0.27	0.95	0.00
pt7	0.25	0.69	0.00
pt8	0.27	0.64	0.33
pt9	0.21	0.77	0.00

Figure 11: SEM surface layer image after the alloying with the electrode of alloy steel 1H18N9 (I = 32 A, Ti = 25 μs); measuring points (Table 2) have been marked.

The points for which the chemical composition has been determined, are shown in Figure 11 with surface layer visible in the background.

In the typical microhardness pattern being a function of distance from the edge, one can see the increase from 600 μHV in the core up to 800–1100 μHV nearby the specimen surface (Figure 12). Microhardness is usually highest at the surface and it gradually lowers with the depth. Micro-hardness of the heat-affected zone is lower than the micro-hardness of the remelted layer but it is higher that the micro-hardness of the material core.

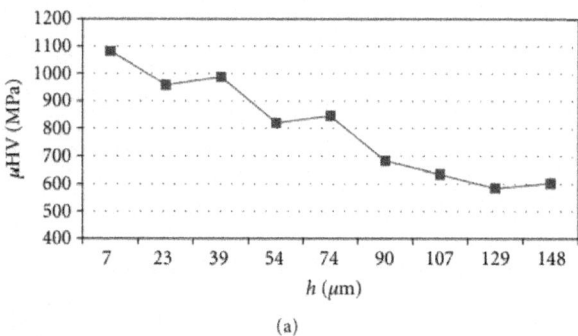

(a)

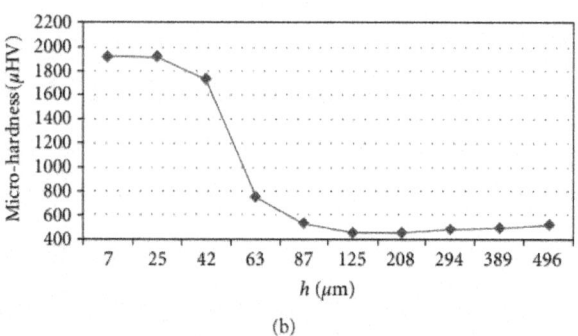

(b)

Figure 12: (a) Microhardness versus the distance from the edge for the specimen machined by the tungsten electrode at the following parameters. (a) EDA I = 24 (A), Ti = 400 (μs), (b) EDA I = 24 (A). Ti = 400 (μs) and then the roto-peen n = 4100 (rev/min), t = 75 (s).

The geometrical microstructure investigation showed that application of the roto-peen after the EDA results in twofold decrease of the roughness, for example Ra goes down from 8 to 4 μm. The obtained micro-structure is of highly favorable pattern (similar to one obtained after the EDM + roto-peen) of flat summits and indents remaining after the EDM (Figures 13 and 14).

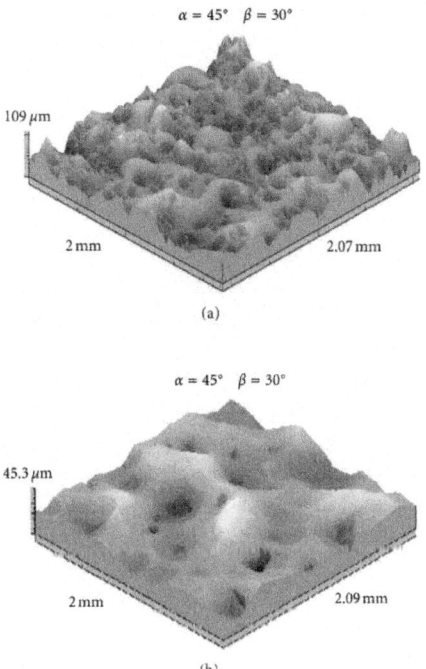

Figure 13: Stereometric image of the surface subjected to the: (a) EDA: 1H18N9 electrode, I = 24 (A), Ti = 400 (μs), U = 200 (V), (c) EDA + roto peen n = 4100 (obr/min), t = 75 (s).

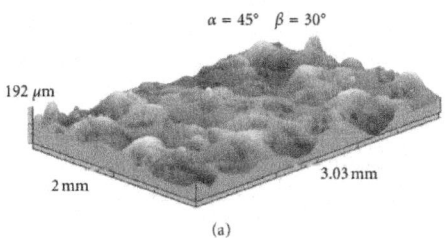

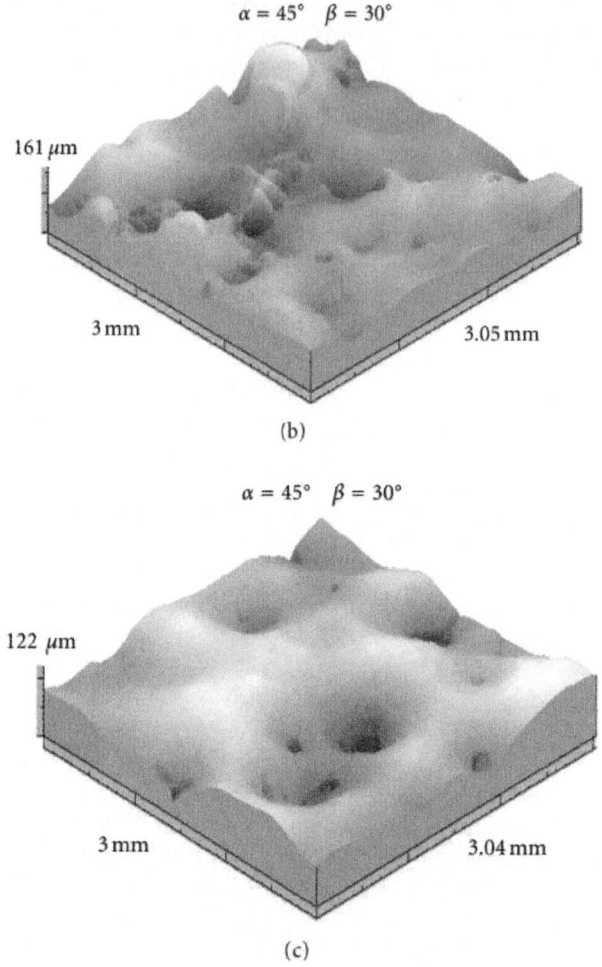

Figure 14: Stereometric image of the surface subjected to the: (a) EDM: cooper electrode, I = 32 (A), Ti = 800 (μs); (b) EDA: 1H18N9 electrode, I = 32 (A), Ti = 3200 (μs); (c) EDA + roto peen; n = 4100 (rev/min), t = 75 (s).

The investigation of performance properties resulting from the EDA and the combined EDM + roto-peen included resistance to abrasive wear and fatigue strength. Figures 15 and 16 show dependence of the observed wear examined by the Skoda-savin method for the lubricated machined samples on the applied manufacturing variants. The performed analysis shows that the EDA application results in considerable resistance to wear when compared to ground surfaces and additional application of the roto-peen after the EDA causes an extra increase in abrasive wear resistance.

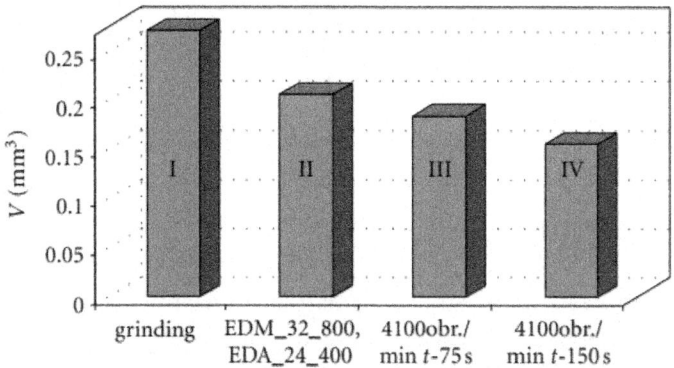

Figure 15: Comparative results. Investigation of abrasive wear resistance measured with Skoda-Savin method: (I) samples of steel 45 machined by grinding, (II) EDM U = 180 (V), I = 32 (A), Ti = 800 (μs) and EDA: U = 200 (V), I = 24 (A), Ti = 400 (μs); (III, IV) EDM, EDA and with roto peen: n = 4100 (rev/min) for the time of: t = 75 (s) (III); t = 150 (s) (IV). (lubrication free conditions– specimen immersion in potassium dichromate).

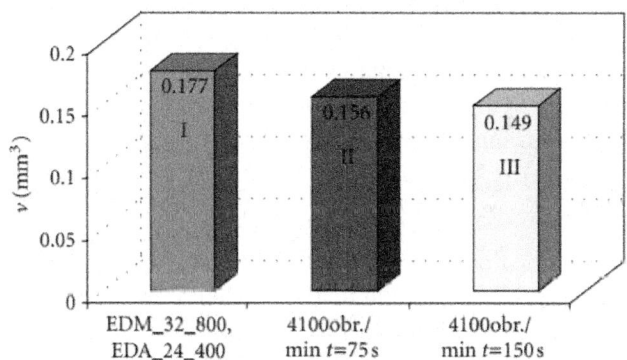

Figure 16: Comparative results. Investigation of abrasive wear resistance measured with Skoda-Savin method: (I) samples of steel 45 machined by grinding, (II) EDM— U = 180 V, I = 32 A, Ti = 800 μs and EDA: U = 200 V, I = 24 A, Ti = 400 μs; (III, IV) EDM, EDA and with roto peen: n = 4100 rev/min for the time of: t = 75 s (III); t = 150 s (IV). (Lubrication conditions—kerosene and machine oil solution −1 : 3).

Preliminary investigation of fatigue resistance of the specimens machined by the EDA + roto-peen, carried out in identical environment like the samples after the EDM showed that the application of the roto-peen machining after the EDA, gives the similar result of fatigue strength increase and durability to the one obtained when applying the roto-peen after the EDM.

CONCLUSIONS

The application of the superficial cold-work treatment represented by the roto-peen machining applied after the EDM and the EDA contributes to considerable lowering of roughness height, (50–70%), reduction in surface layer microcrack and discontinuity number as well as elevation of surface layer microhardness by 300 µHV in the case of the EDM and 700 µHV for the EDA.

The roto-peen application results in considerable increase of abrasive wear resistance, 3-fold for the EDM, and 40% for the EDA.

The roto-peen application is relatively cheap and practically simple operation which can be realized with the universally available equipment; both the EDA and the roto-peen machining are particularly dedicated to the treatment of small surface areas or the selected portions of large parts.

The EDA and roto-peen machining can be applied for improving performance properties of machine and tool parts which are the most exposed to wear.

REFERENCES

1. J. P. Kurth, J. Van Humbeeck, and L. Stevens, "Micro structural investigation and metallographic analysis of the white layer of a surface machined by Electro Discharge Machining," in Proceedings of the 11th International Symposium on Emergency Management (ISEM '95), pp. 849–862, Lausanne, Switzerland, 1995.

2. M. Siwczyk, Electrical Discharge Machining Obróbka Elektroerozyjna, WNT, Warsaw, Poland, 1981.

3. B. Nowicki, R. Pierzynowski, and S. Spadło, "Comperative investigation into the brush electrodischarge alloying with the electrodes of alloy steel and tungsten," International Journal for Manufacturing Science and Technology, vol. 4, no. 1, pp. 44–54, 2004. View at Google Scholar

4. Collective work Electrical discharge allying of metal surfaces, Elektroiskrowoe legirowanie metaliceskich powerchnostej, Akadamia Nauk USSR, 1976.

5. L. C. Lim, L. C. Lee, Y. S. Wong, and H. H. Lu, "Solidification microstructure of electrodischarge machined surfaces of tool steels," Materials Science and Technology, vol. 7, no. 3, pp. 239–248, 1991.

6. J. A. McGeough and H. Rasmussen, "A macroscopic model of electro-discharge machining,"International Journal of Machine Tool Design and Research, vol. 22, no. 4, pp. 333–339, 1982.

7.   A. Dmowska, B. Nowicki, and A. Podolak-Lejtas, "A new method of investigating crater and flash made by individual discharge using scanning profilometers," Metrology and Properties of Engineering Surfaces, Rzeszów, pp. 139–145, 2011.

8.   B. Ekmekci, "Residual stresses and white layer in electric discharge machining (EDM)," Applied Surface Science, vol. 253, no. 23, pp. 9234–9240, 2007.

9.   H. T. Lee and T. Y. Tai, "Relationship between EDM parameters and surface crack formation," Journal of Materials Processing Technology, vol. 142, no. 3, pp. 676–683, 2003.

10. P. G. Bailey, "Manual peening with the rotary flap process," in Proceedings of the 7th International conference on shot peening, Poland, pp. 405–414, Warsaw, Poland, 1999.

11. M. Drozd, A. Fedorov, and J. Cidjakin, "Rascet głubiny rosprostranienia plasticeskih deformatcji,"Westnik Mashinostroenija, no. 1, pp. 132–137, 1972.

12. A. Nakonieczny, T. Żółciak, and G. Monka, "Effects of shot peening on process of carburization and selected strength properties of steel 18 HGT," in Proceedings of the International Conference on Shot Peening, pp. 135–144, Warsaw, Poland, 1999.

13. W. Przybylski, Burnishing Technology Technologia Obróbki Nagniataniem, Wydawnictwo Naukowo-Techniczne, Warsaw, Poland, 1987.

14. Y. C. Lin, B. H. Yan, and F. Y. Huang, "Surface improvement using a combination of electrical discharge machining with ball burnish machining based on the taguchi method," International Journal of Advanced Manufacturing Technology, vol. 18, no. 9, pp. 673–682, 2001.

15. B. Nowicki and A. Podolak-Lejtas, "Investigations of the effect of combined EDM machining with burnishing process on the conditio of the surface layer," Advances in Manufacturing Science and Technology, vol. 32, no. 8, 2008.

16. American National Standard for Surface Integrity B 211.1.1986.

Chapter 10

RECENT DEVELOPMENTS AND RESEARCH ISSUES IN MICROULTRASONIC MACHINING

Vivek Jain Apurbba Kumar Sharma and Pradeep Kumar

Mechanical and Industrial Engineering Department, Indian Institute of Technology Roorkee, Roorkee, Uttarakhand 247667, India

ABSTRACT

Demand for micromachining has been on the rise in recent years owing to increasing miniaturization. Production of parts in microscale, especially with brittle materials, is challenging. Ultrasonic micromachining has been gaining popularity as a new alternative in fabrication of such parts. The process gives a machining option for geometrically challenging and/or brittle material parts that are difficult to machine by conventional processes. In the recent years, possibilities have been explored to improve the "Unit Removal" in microultrasonic machining (micro-USM). However, the research in the area is yet to attain momentum. The present paper is an attempt to present the state of the art in the area of micro-USM based on the literature. Developments in the critical areas of the process like machine tool technology, machining tool head, transducers, and precision attainable in the process with challenges have been discussed. Potential research issues have been explored for future work. Possible application areas have been identified.

INTRODUCTION

Miniaturization has been one of the major themes in manufacturing in the recent years after "automation" took the researches and industries the same way during the last quarter of the previous century. Research publications in the area of micromachining started appearing since the end of the last decade, and in fact, got the momentum with the unfolding of the new century, mostly driven by industry needs. Miniaturized components and products have been in strong demand because of their unique advantages such as less space requirement, less energy and material consumption, easeness in carrying, and/or handling and can be cheaper. The growth of Micro Electro Mechanical Systems (MEMS) and the related research in different industries such as electronics, optics, medical, biotechnology, automotive, communications, and avionics are largely attributed to these microcomponents. But product miniaturization poses multiple challenges and demands innovations as well as continuous improvements in manufacturing technologies to ensure the processing of a wide range of materials [1]. Hence micromachining is considered as one of the key enablers to solve these issues and consequently, in great demand. Research is active in this area to characterise new technologies or to improve known processes to guarantee fine precisions and low costs of manufacturing, so as to facilitate a real growth and diffusion in the industrial world. Several research issues associated with micromachining starting from standard lithographic methods to nontraditional technologies and mechanical micromachining techniques have been reported [2]. Bulk micromachining, which utilizes deep etching techniques, was developed and used to produce pressure sensors, accelerometers and ink nozzles [3]. The basic scheme for surface micromachining was developed and introduced in 1983 [4]. In the 1980s, the LIGA (Lithographie Galvanoformung Abformung) process was developed in Germany [5] and became popular because it can be used to make parts or moulds from electroplatable materials or use the moulds to make injection moulded plastics. While these techniques can be used for mass production in microchip manufacturing, industrial applications for complex three-dimensional systems in microscale are still limited by materials, geometrical limits, and fabrication process costs and time. Bulk micromachining has the disadvantage that the devices are generally relatively large and therefore consume most of the chip area [6]. In regard to LIGA, minor precisions and aspect ratios are the limits of the process.

Tool-based micromachining methods such as microturning, micromilling, and microgrinding are capable of generating 3D free-form features with micron level accuracy. Despite its potential, practical use of mechanical microcutting is limited by the low tool/machine stiffness and cutting tool strength, especially for hard materials such as heat-treated mold, die steels, and ceramics [7, 8]. One solution to this problem is to employ nontraditional machining techniques which are also being developed in microscopic scale in order to achieve real 3D sculptured complex components. Interesting methods are, in particular, laser micromachining, micro-electrodischarge machining (micro-EDM), micro-electrochemical machining (micro-ECM), and micro-USM.

Glass, silicon, and polymers are the most common micromachined materials using lasers [9, 10]. Ink jet printer nozzles are an example of devices manufactured by excimer laser micromachining [11]. Micro-electrodischarge machining is commonly used to microdrill or to pattern trenches in workpieces, with a micromilling approach and simple-shaped tools [12, 13]. Circular rotating micro-tools are mainly adopted and on-the-machine made in order to avoid off-centering and tilting errors. Further, the problem of imprecision associated with tool rechucking on tool holder is significantly minimized, improving at the same time roundness quality during machining. The area is strengthened by the emergence of micro-electrochemical machining that appears to be a very promising micromachining technology due to its advantages that include high MRR and rapid machining time. It also permits machining of chemically resistant materials like titanium, copper alloys, super alloys, and stainless steel, which are widely used in biomedical, electronic, and MEMS applications [14]. Despite the significant benefits of these techniques, important issues like (i) the presence of a heat affected zone (HAZ) and (ii) thermal stresses in the micromachined part in case of laser micromachining [15] and micro-EDM are yet to be solved. Machining time and material removal rates are mainly limited by small discharge energies in micro-EDM [16]. In micro ECM, on the other hand, the dissolution rate of the metal workpiece decreases as the machining depth increases on account of the difficulty of maintaining the flow of electrolyte deep inside the microhole [17]. The flow space of the electrolyte and electrolyte diffusion was improved via the application of ultrasonic vibrations. Some improvements in machining efficiency are also obtainable with ultrasonic tool or workpiece vibration while superimposed during micro-EDM drilling operations, in relation to the better removal of debris [18, 19]. Since ultrasonic vibration complements other nontraditional processes, special attention is required to the role of microultrasonic machining. Thus, an attempt has been made to present a detailed review of the developments in the area of micro-USM process and various research issues involved in making the process more pragmatic. Significantly less published data in the area of micro-USM indicates that the area is yet to be well explored. However, research activities in the area have been increasing in the current decade.

MICRO-USM PROCESS

Micro-USM is basically evolved from macro-USM, which has been already investigated and reported widely. Micro-USM as one of the nontraditional manufacturing processes finds its main advantage in machining nonconductive, hard, and brittle materials and capability of generating surface free of thermal damage. Table 1 presents a comparison between different micromachining techniques. The micro-USM process, as observed from the table, can be compared to almost all other micromachining processes. This process has the edge over the highly competitive LIGA process as far as 3D profile machining is concerned apart from applicable to a broad range of work materials.

Table 1: Common attributes of different micromachining techniques [2, 3, 8]

Features	Micromachining techniques						
	Bulk-surface	LIGA	Laser	Micro-EDM	Micro-ECM	Micromilling	Micro-USM
Minimum Dimension	+	++	++	++	+	+	++
Accuracy	+	++	±	+	+	±	+
Aspect ratio	–	++	±	±	±	±	±
MRR	Depends upon process	Depends upon process	+	++	+	+	+
Thermal damage to the surface	No	No	Yes	Yes	Yes	Yes	No
Geometrical freedom	2D	2D	3D	3D	3D	3D	3D
Material	Very limited material suite	Very limited material suite	Metal, polymers, ceramics	Only conductive materials	Only conductive materials	Metal, polymer	Conductive, nonconductive, hard, and brittle materials

++ Very good; + good; ± fair.

Historical Development

Ultrasonic machining is a relatively old machining technique whose basis was laid way back in 1927 and was patented by L. Balamuth in 1945. The technology that has evolved since then has been variously termed "ultrasonic machining", "impact grinding", or "slurry drilling" and relies on the cutting action of an abrasive slurry flowing between the vibrating tip of a transducer and a workpiece. It is a technology that has attained a recognized status in manufacturing technology and found increasing applications in industries including aerospace, optics, and automotive [20]. The first USM tools, mostly mounted on the bodies of drilling and milling machines, had been built by 1953-1954. By 1960, independent USM tools of various types were commercialised and came into regular production for a variety of applications [21]. The first attempt of downsizing macro-USM for micromachining was conducted by Masuzawa of Tokyo University in the mid 1990s [22]. The micro-USM is used for machining hard and brittle materials. Typical workpiece materials being machined in previous experimental investigations include glass, silicon, and alumina [23, 24]. Micro-USM has a set of process parameters similar to macro-USM. However, the downsizing for micromachining requires a microsized tool (or tool feature), smaller amplitude, and microsized abrasive particles. A comparison of the micro-USM and macro-USM parameters is presented in Table 2.

Table 2: Features of macro- and micro-USM

Parameters	Micro-USM	Macro-USM
Vibration frequency, kHz	Usually >20	Usually >20
Vibrated part	Tool or Workpiece	Tool
Amplitude, μm	Within microns (0.5~5)	Tens of microns (8~30)
Abrasive particle size, μm	Within microns (0.5~5)	Tens of microns (50~300)
Static load	Gram force	Kilogram force
Tool/feature Size, mm (End diameter)	Within 0.5	Usually >1

Process Principle

Figure 1 illustrates a basic setup of a micro-USM process. The set up primarily consists of a tool system and slurry supply unit. The micro-USM employs the mechanical vibration of the ultrasound with a frequency within the range of 20 to 40 kHz. The tool is mechanically vibrated at an ultrasonic frequency and amplitude of few micrometers. The abrasive slurry, a mixture of irregular-shaped fine abrasive particles (usually, in the range of (0.5~5 μm)), and a liquid medium is fed into the gap between the tool and workpiece. As the vibrating tool head hits the free abrasives in the slurry, they attain momentum and impact upon the target workpiece location. A localized fatigue stress is developed in the impact zone owing to continued impact, and microchipping occurs resulting in material removal. Moreover, a small amount of material removal might also be contributed by the mechanical abrasion of the hard microabrasives. Further, implosion of the gas bubbles, also called cavitation, can play a key role in material removal at microlevel. Although water is usually preferred as the slurry medium, there exists a possibility that the chemical impurities present in the slurry medium can selectively cause instantaneous degradation of the work material resulting in loss of material. A continuous flow of slurry flushes away the debris from the machining zone and refills the gap with fresh slurry.

Thus, based on the understanding of macro-USM research, the contributing mechanisms in case of micro-USM can be summarized into four categories:

- microchipping by impact of the free moving abrasive particles,
- mechanical abrasion by the abrasive particles against the workpiece surface,
- cavitation effects in liquid agitated by ultrasonic vibration,
- chemical actions associated with the liquid being employed.

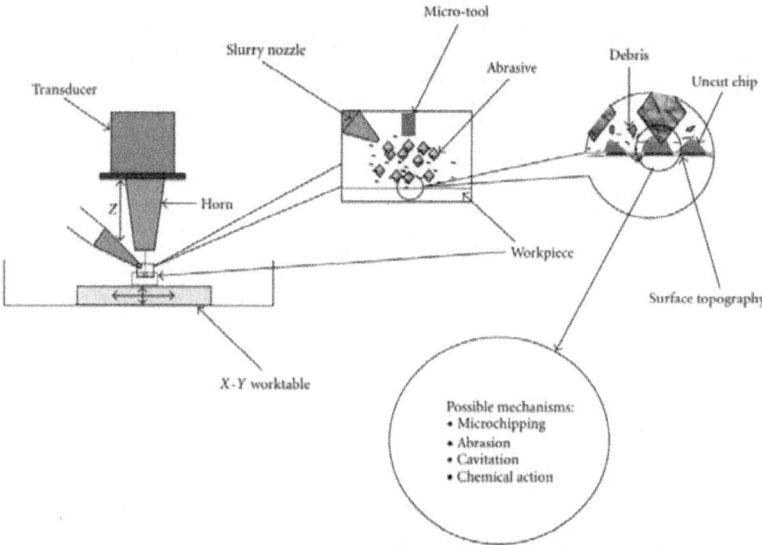

Figure 1: Principles of micro-USM.

DEVELOPMENTS IN MICRO-USM MACHINE TOOLS AND TOOL HEAD

The variants in micro-USM can be categorised as shown in Figure 2. While Figure 2(a) presents a classification based on the machine tool characteristics, Figure 2(b) shows some variations in the micro-USM according to different tool heads used. The first type of micro-USM originates from the concept of "micromachining using microtool" [24, 25] in which there is no rotary motion involved. A simple cylindrical tool was used, and the vibration was given to the microtool. As USM is associated with a major drawback of tool wear, this kind of tooling causes continuous shortening of tool length and, therefore, imposes obstacles in maintaining consistent vibration amplitude at the tool tip. The vibration amplitude varies at different locations along the tool axis, and tool wear changes the location of the tool tip causing the inconsistency. Applying ultrasonic vibration to the workpiece has been found to be preferable because it eliminates the influence of tool wear on the vibration amplitude of tool tip in case of applying vibration to the tool. Furthermore, the vibrated workpiece may help in stirring the abrasive slurry during machining to increase the efficiency of abrasive particles around the machining zone and remove debris [26].

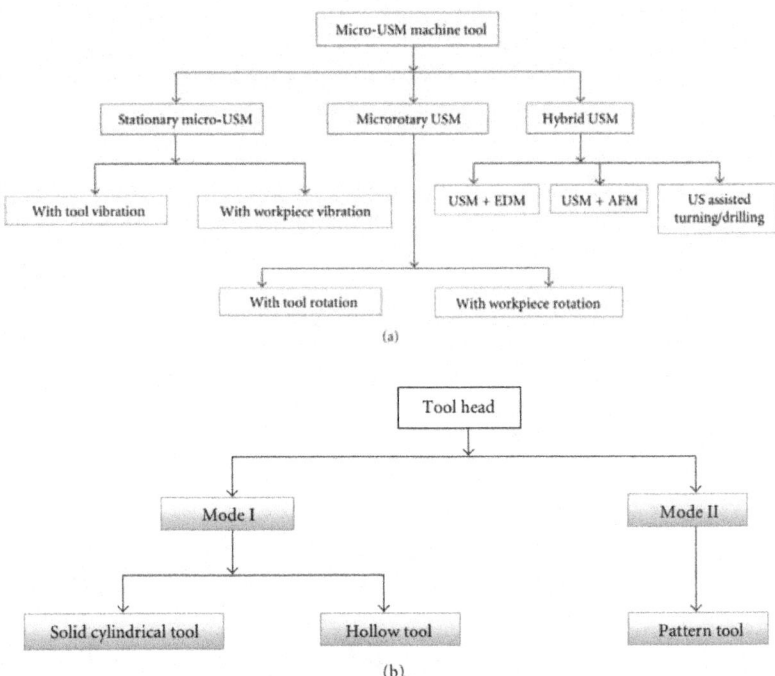

(a)

(b)

Figure 2: Variants of micro-USM based on (a) machine tool characteristics and (b) different tool heads.

Rotary ultrasonic machining (RUM), as indicated in Figure 2(a), is one of the cost-effective and hybrid machining processes available for drilling holes. It merges the material removal mechanisms of diamond grinding and USM, resulting in higher material removal rate (MRR) than that obtained by either diamond grinding or USM [27]. Figure 3 presents a schematic illustration of RUM. The hollow tool can be given a rotary motion as well as it can vibrate. But this makes the design very unwieldy. Hence, an attempt has been made to study the material removal rate when the workpiece is rotated in rotary USM mode. However, owing to the rotation of the workpiece there will be sliding and rolling contacts between the abrasive grains and workpiece, as well as impacts and indentation of the abrasive particle with the workpiece at ultrasonic frequency. Hence better material removal has been claimed [28]. Two major requirements for micro-RUM are the microsized abrasive bonded tool and a machining system capable of applying very small load on the microtool with necessary feedback and control mechanisms.

Some of the disadvantages of the EDM, ECM, and other nonconventional machining processes were overcome by ultrasonic assistance in the form of hybridisation (Figure 2(a)). Thus, ultrasonic machining was combined with

EDM and abrasive flow machining (AFM) to achieve better yield [17–19, 29]. Nowadays, ultrasonic vibrations are being used successfully to enhance machining capability of micro-EDM to handle titanium alloys [30]. It has been found in microhole machining of titanium plate that microultrasonic vibration lapping enhances the precision of microholes drilled by micro-electrodischarge machining. Further, USM assisted turning is claimed to reduce machining time, workpiece residual stresses, and strain hardening and improve workpiece surface quality and tool life compared to conventional turning [31].

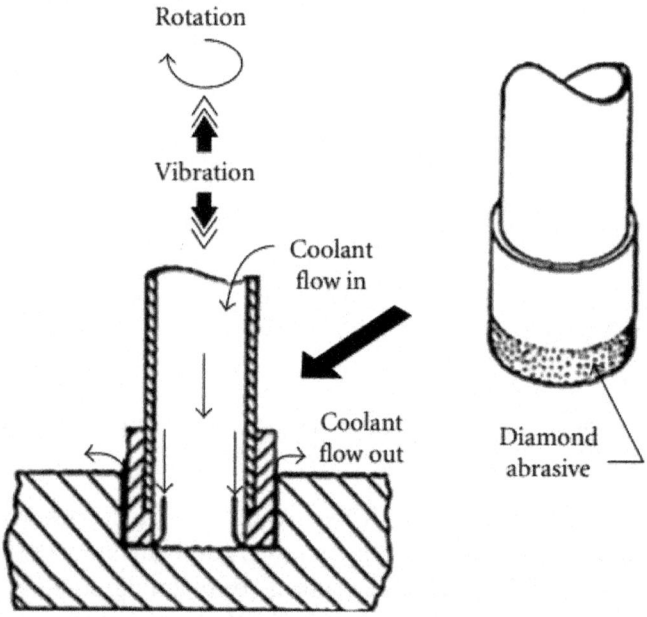

Figure 3: Rotary ultrasonic machining process [27].

On the basis of tool head, micro-USM can be divided into two major types: mode I and mode II (Figure 2(b)). Mode I consists of a solid or hollow cylindrical tool which is easy to fabricate and widely used in various applications of micro-USM. Figure 4 illustrates the mode I tool. In mode II, microfeatures are fabricated on the tool bottom. The tool functions as a pattern and is travelled vertically toward workpiece, and, therefore, microfeatures can be "replicated" onto the workpiece in one sinking operation. A unique benefit of this type of micro-USM is to realize a parallel production of many identical simple features. The tool is silver brazed with the tool head and not touching the workpiece. Figure 5 shows the microfeature developed on the tool itself and a gang drill.

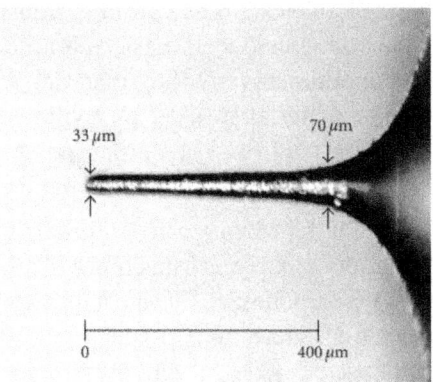

Figure 4: A typical microtool [32].

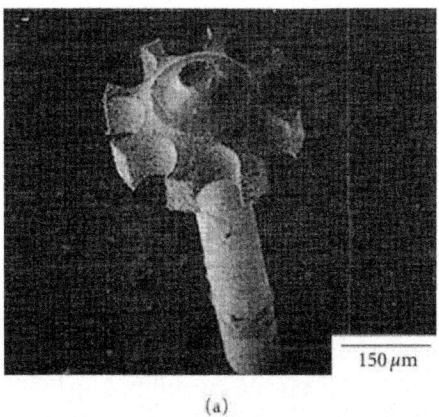

(a)

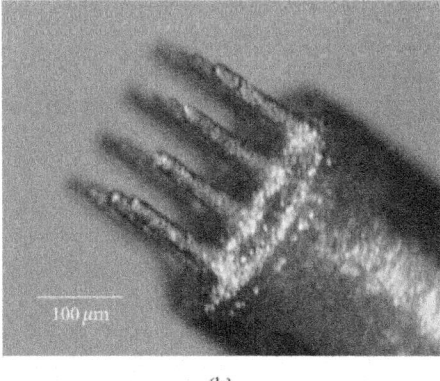

(b)

Figure 5: Scanning electron micrograph of a (a) 3D microgear and (b) gang drill produced through microfeatures [22, 34].

In order to minimize tool wear, tools should be constructed from relatively ductile materials such as stainless steel, brass, and mild steel. Curodeau et al. (2008) have proposed an alternative tooling material, in which a viscoelastic thermoplastic composite material is used as tooling to conduct ultrasonic micromachining operations [33]. The proposed tool was successfully investigated for micromachining and micropolishing for tool steel surface.

The machine tools for USM range from small, tabletop-sized units to large-capacity machine tools. All USM machines share common subsystems regardless of the physical size or power. The most important of these subsystems are the power supply, transducer, tool holder, tool, and abrasives [35]. In the case of USM transducer, electrical energy is converted into mechanical motion. With a conventional generator system, the tool and horn are set up and mechanically tuned by adjusting their dimensions to achieve resonance. Recently, however, resonance following generators has become available which automatically adjust the output high frequency to match the exact resonance of the horn/tool assembly. They can also accommodate any small error in setup and tool wear, giving minimum acoustic energy loss and very small heat generation. The power supply depends on the size of the transducer.

Two different types of transducers used for USM work on two different principles of operation-piezoelectric and magnetostrictive. Piezoelectric transducers generate mechanical motion through the piezoelectric effect using certain materials such as quartz or lead zirconate. Piezoelectric transducers, by nature, exhibit extremely high electromechanical conversion efficiency (up to 96%), which eliminates the need for the water-cooling of the transducer. These transducers are available with power capabilities up to 900 W. Magnetostrictive transducers are usually constructed from a laminated stack of nickel or nickel alloy sheets. These types of transducers are rugged but have electromechanical conversion efficiencies ranging from only 20 to 35% [36].

MICRO-USM PERFORMANCE MEASURES

Extensive research work has been carried out for macro-USM regarding material removal, tool wear, dimensional accuracy, and surface quality [37–39]. However, only a few publications have focussed on the ways of yielding optimal micro-USM performance measures in terms of high material removal rate (MRR), low tool wear rate, and satisfactory surface quality.

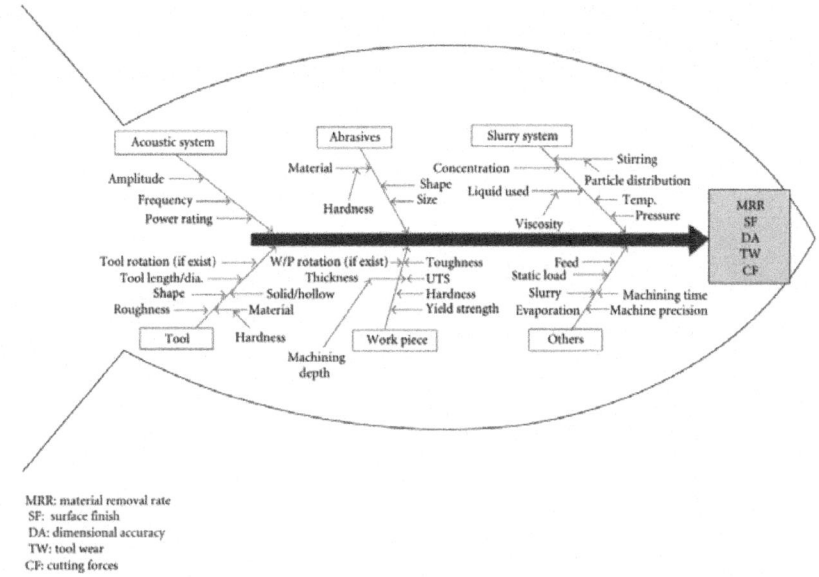

MRR: material removal rate
SF: surface finish
DA: dimensional accuracy
TW: tool wear
CF: cutting forces

Figure 6: Subsystems and associated parameters in micro-USM.

The main task of process improvement for micro-USM is to economically machine a microfeature with required surface topography, minimal surface damage, good surface finish, high-dimensional accuracy, and acceptable material removal rate. The parametric relationship for micro-USM is complicated due to the involvement of numerous factors and related parameters that could affect the process outputs. The possible parameters in micro-USM can be illustrated in a fishbone diagram as shown in Figure 6.

Effect of Operating Parameters on Machining Rate

Machining characteristics of micro-USM in the dimension range under 100 μm have been first reported and discussed by Egashira et al. [25]. The type and size of abrasive particle affect the machining rate significantly. The machining rate increases with the size of abrasives. The machining speed increases with an increase in the average static load; however, it decreases with the increase of the average static load beyond a certain value. The debris accumulation in the working area leads to a part of the static load consumed in impacting the debris instead of removing the material from the workpiece, resulting in a lower machining efficiency [40, 41]. The rotation of the tool improves the machining speed significantly as shown in Figure 7. The tool rotation helps in the debris removal and, therefore, increases the machining speed. The drilling speed, however, was reported to be rising with increase in the machining load as that for a single tool does while using cemented carbide tools. In contrast, it decreases as the oscillation amplitude increases [34]. The effects of oscillation amplitude and machining load on drilling speed and tool wear ratio, as reported by Egashira et al. (2004), are illustrated in Figure 8.

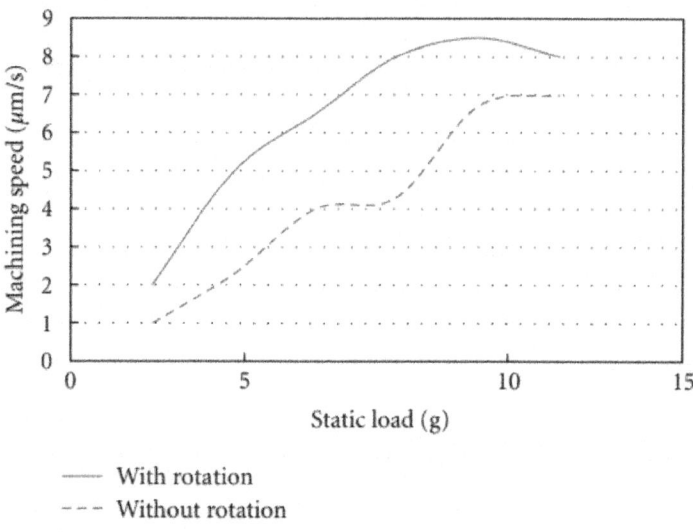

Figure 7: Machining speed as a function of average static load (based on data [40]).

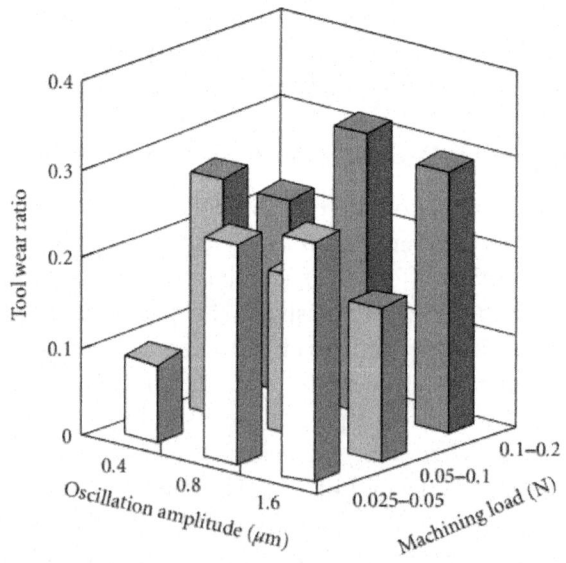

Figure 8: Tool wear ratio for cemented carbide tool [34].

Effect of Operating Parameters on Tool Wear

In USM, the grains hit the tip of the vibrating tool and tend to erode it. Thus, tool wear is an important variable for micro-USM, affecting the machining speed and the hole accuracy. Tool wear tends to increase when harder and coarser abrasive grains are used [25]. When tools of very small dimensions are used, the static load needs to be small to avoid breakage of the tool. A load between 50 to 100 g for a tool with 65 μm diameter or a load of 10 g for a 20 μm² square tool seems to be the best values as reported by Boy et al. (2010) [42]. The populous tool materials such as stainless steel and thoriated tungsten in conventional USM are not suitable in micro-USM because of their large wear. Also, tool wear increases remarkably with a decrease in tool diameter [22, 43]. The tool wear ratio is defined as the ratio of the tool wear length to the hole depth. The tool wear ratio for cemented carbide tools varies widely and is influenced by the machining load or oscillation amplitude as illustrated in Figure 8 [34]. However, larger amplitude than 2 μm or a heavier load than 5 mN often results in tool breakage [26].

Effect of Tool Geometry on Tool Wear Ratio and Machining Rate

Some studies on possible effect of tool geometries on machining rate and tool life were also carried out by the present authors. Measurements of machining rate and tool wear ratio as a function of drilling depth were performed using hollow and solid tools. Tools made up of austenitic stainless steel and the hollow tools were having a circular cross section of 500 μm outside diameter and 450 μm inside diameter. The silicon wafers with 500 μm thickness were employed as workpiece material for the drilling trials. The experiments were carried out in a stationary Sonic-Mill machine (AP-500) with a power output of 500 W. A 20% power rating was used during the experiments. The static load applied on the horn was taken in the range of 100 g to 300 g throughout the trials.

Figure 9 illustrates the variations in the observed machining rate (MR) as a function of the static load for the two types of tool. In both types of tool used, the overall machining rate increases with an increase in the static load. Increase in static load means increasing the pressure on the abrasive grains and eventually on the workpiece, resulting in an increase of machining rate. However, the machining rate decreases with the increase in static load differently for the two types of tool. Decrease in MR with increasing static load is explained by the insufficiency in recycling abrasive particles at the machining interface because of accumulation of the debris. The recycling capacity affects the hammering and impact actions of the abrasive particles in the working gap. Also, the MR is more in case of hollow tool with the varying static load. This is because of the necessary contact area between the tool and abrasives and correspondingly between abrasives and the workpiece. In case of hollow tool, material is removed by the border of the tool only whereas in case of a solid tool, the whole lower edge is involved in machining. The design of a tool must be such that it should machine as little of the workpiece material as necessary. The importance of tool head area can also be seen from the accompanying scanning electron micrograph (SEM) images in Figure 9. Higher straying abrasive action is seen with the solid tool, while the hollow tool provides a more focused cutting zone, although at the cost of higher tool wear. Thus, further studies in this direction would be needed to obtain optimal machining conditions.

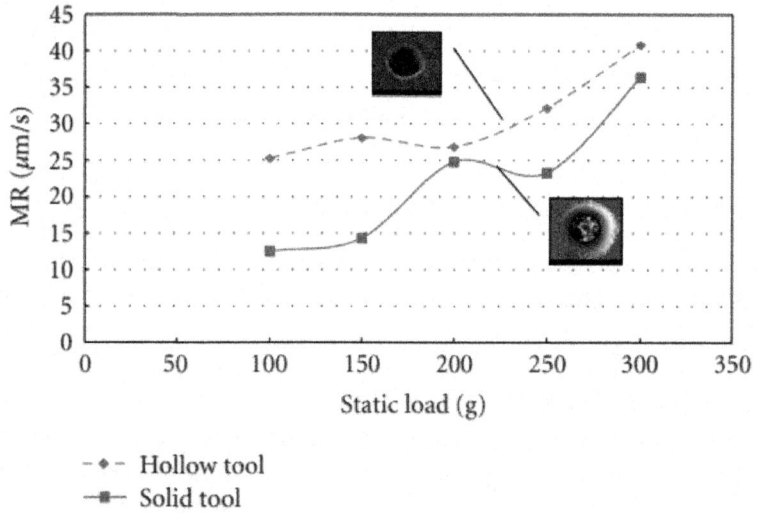

Figure 9: Variation of machining rate as a function of static load; insets: SEM micrographs of ultrasonically machined surfaces of silicon with hollow tool and solid tool.

Figure 10 illustrates the tool wear ratio of different types of tool used. The observed higher tool wear ratio in case of hollow tool is attributed to reduced contact area. As the contact area is less for a constant load, the stress produced will be more which results in easy and quick work hardening of the tool tip. This leads to an induced brittleness on the tool tip causing a favourable condition to be eroded by the deflected abrasives. In addition, the problem of debris accumulation also causes the crack generation and material removal from the tool, leading to the more tool wear.

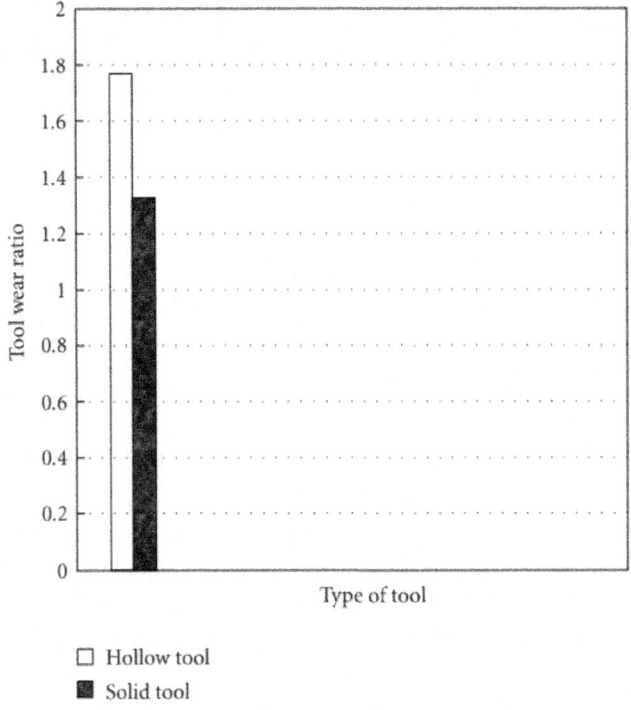

Figure 10: Tool wear ratio of different tool.

Effect of Operating Parameters on Surface Finish

The type and size of abrasive particle affect the surface finish and it increases as the size of abrasive increases. Egashira et al. have reported that tungsten carbide (WC) causes the worst surface as compared to titanium carbide (TiC) [25]. Theoretically, the surface roughness is small when small particles are used because of the small crater generated. However, because of debris accumulation, the surface roughness generated by small abrasive particles can be larger than that of big particles. If abrasive particles of small size are used, particles get embedded in the workpiece or in the tool under the static load and may stay in their original craters without moving to other locations during several cycles of vibration. The thickness of accumulated debris in the working area reaches the exposed height of abrasive particles in a short time, resulting in the debris involved in machining. This leads to a change in the abrasive particle distribution and movement, which may increase the surface roughness. When large abrasive particles are used, more debris is removed from working area and the abrasive particles easily move to new locations by the vibration because of the large number of impacts [40,41]. Surface roughness, out-of-roundness and taper ratio are three important parameters for evaluating microhole quality [22, 43]. Cutting forces are reduced by 60 to 70% with ultrasonic vibrations which result in increase in tool life, penetration, and tool length, thereby improving surface finish, and machinable depth [44]. Rotating the tool not only decreases the severity of protrusions along the machining path, but also improves machining speed as observed by Kuriyagawa et al. (2001) [32].

GEOMETRICAL CAPABILITIES

The geometrical capabilities of micro-USM have been verified by machining microfeatures such as blind or through holes, slots, and 3D cavities. Ultrasonic machining has a limitation in its application to micromachining because there are problems in fixing microtools to the machine and maintaining high precision. Accordingly, a technique was proposed for micro-USM by applying on-the-machine tool fabrication by wire electrodischarge grinding (WEDG). As a result, microholes as small as 20 μm in diameter and 50 μm depth on a silicon plate, and quartz could be made. Figure 11 shows a triangular and square hole made on silicon using WEDG method, as reported by Egashira et al. (1997) [25].

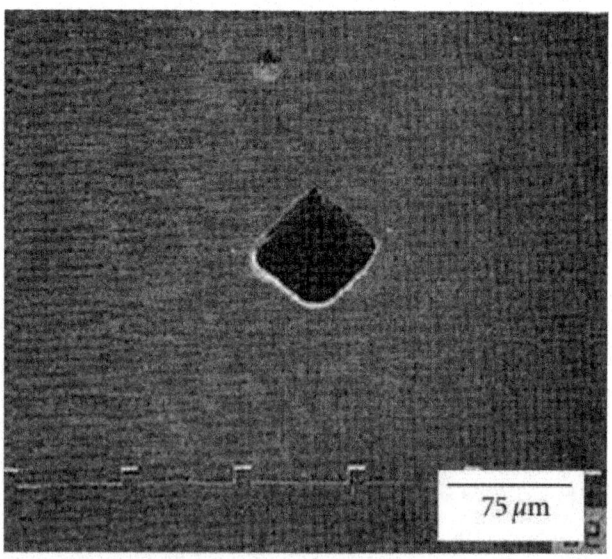

(a)

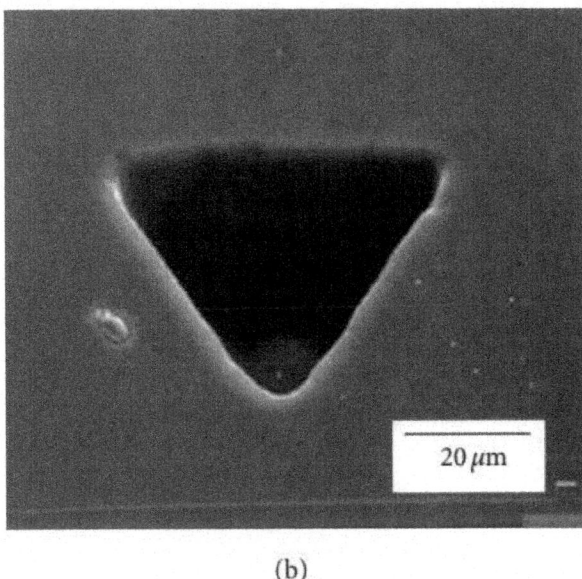

(b)

Figure 11: Square and triangular holes on silicon [25].

A machine was developed at FEMTO-ST institute, France for producing electronic components based on piezoelectric quartz crystals for ultraprecise processing on 2-in. wafer. Through holes of high quality were easily produced; blind holes were also machined with a depth accuracy within 10 μm. Figure 12 illustrates arrays of pillars of 280 μm diameter and 6,000 μm depth in Lead Zirconate Titanate (PZT), which were produced using a steel disk with an array of 300 μm diameter holes (honeycomb structure) [42]. A hole of 15 μm in diameter but only 32 μm depth has been achieved by Sun ct al. [22]. An array of microholes were drilled in micro-USM using a cemented carbide multitool having diameter 18 μm as shown in Figure 5(b). The drilling was performed four times and a total of 64 holes could be fabricated with an oscillation amplitude of 0.8 μm and a machining load of 0.05 to 0.1 N. The workpiece material was soda lime glass. In an extreme attempt, micro-USM was used to drill microholes of diameter 5 μm in silicon and soda lime glass as shown in Figure 13.

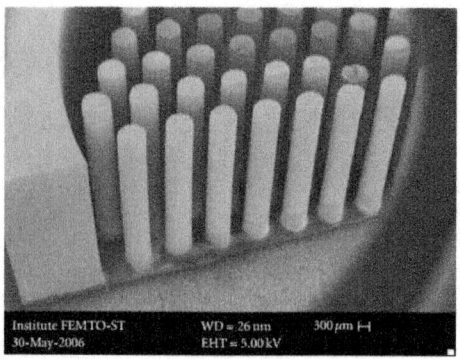

Figure 12: Typical ultrasonically machined structure [42].

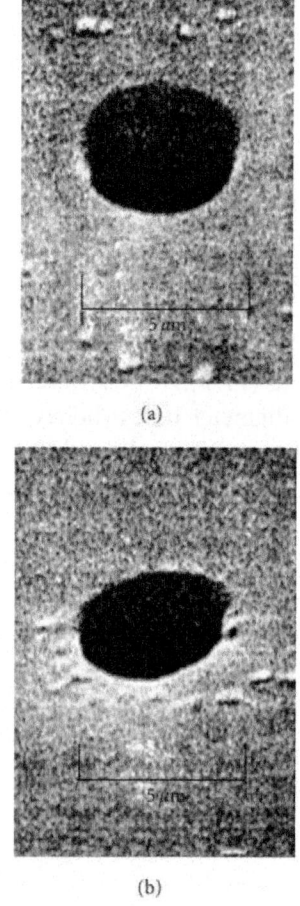

Figure 13: Microholes in silicon and glass [25, 26].

The major efforts to improve the micro-USM process performance are to increase its throughput by using novel microtooling technology, especially for Mode II micro-USM (Figure 2(b)). Micro-USM with a single tool had so far been carried out on hard and brittle materials; however, it is time-consuming for drilling multiple holes, which is often a requirement for application of microholes. Thus, a micro-EDM has been used to drill parallel holes, which is later used to produce microcemented carbide multitool using reverse micro-EDM. The multitool is further used in mode II micro-USM. The fabricated multitool (Figure 5) and the array of microholes machined by micro-USM are shown in Figure 14.

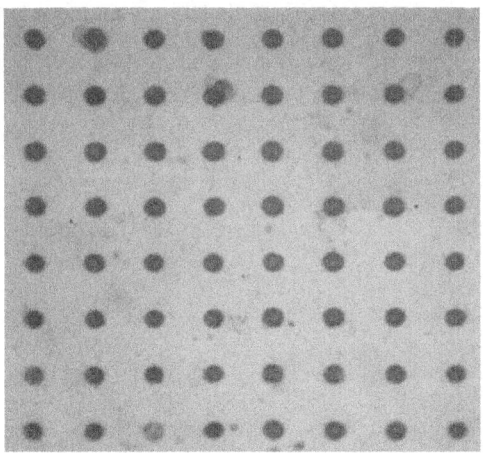

Figure 14: Microholes 20 μm in diameter [34].

A fabrication process combining lithography, electroplating, batch mode micro-electrodischarge machining (μEDM) and batch mode micro-USM, LEEDUS, was developed by Li and Gianchandani (2005) to provide die-scale pattern transfer capability from lithographic masks onto ceramics, glass, or other brittle materials. A die (eventually a wafer) with electroplated metal structures (the desired pattern needs to be present in the final machined part) was used as a micro-EDM electrode. The die-scale (or wafer-scale) negative pattern is transferred onto the tool material during micro-EDM, and then the positive pattern is transferred onto the workpiece by mode II Micro-USM [46].

Figure 15(a) shows a straight slot (width 47 μm, length 500 μm, and depth 60 μm) with vertical side wall being machined by Sun et al. (1996) using a layer-by-layer contouring mode through micro-USM system. Furthermore, contouring mode micro-USM was illustrated by generating a spiral groove on low-melting glass as shown in Figure 15(b) [22].

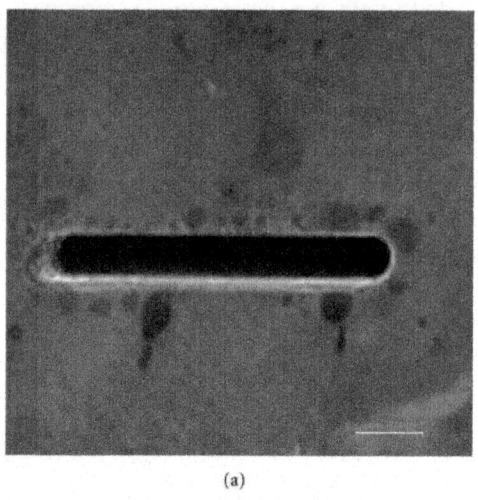

(a)

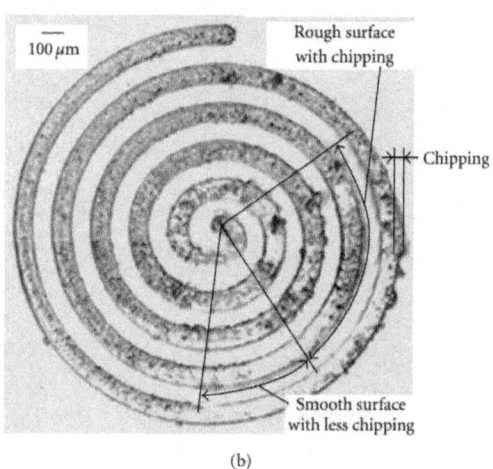

(b)

Figure 15: (a) Straight microslot, (b) Spiral groove [22].

A complex-shaped microfeature can be generated by moving the microcylindrical tool along with a designated tool path. An arbitrary 3D micro cavity was successfully machined on silicon using a cylindrical-shaped microtungsten tool by Yu et al. (2004) as shown in Figure 16 [41]. During machining, the microtool was driven in all three axes to follow a designed tool path created by CAD/CAM method. To compensate the large tool wear, the "Uniform Wear Method", which was originally developed for producing 3D features in micro-EDM, was applied to postprocess a tool path generated by CAD/CAM system. Tool shape was maintained by specifying layer-by-layer machining.

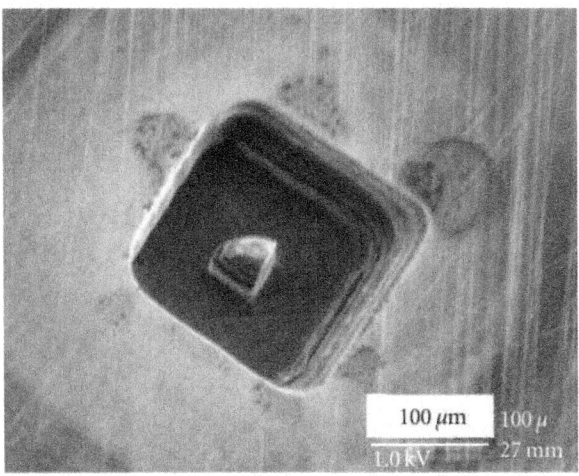

Figure 16: A typical 3D cavity [41].

Figure 17 shows a feature machined on alumina, considered to be an important step in realizing mode II micro-USM for pattern transfer in fabricating complex shaped tool. Such fabrication is obviously more difficult than producing simple cylindrical tool. The newly developed microultrasonic assisted lapping technique shows flexible capability in generating microstructures of various geometrical forms [45]. Major micro-USM process capabilities can be summarised in Table 3. However, there are many technical constraints to be overcome like the achievable concave feature size, aspect ratio, and surface roughness limits, as well as an operational issue incurred by tool wear. A focussed discussion on process capabilities of the micro-USM process was also presented by Hu et al. (2006) elsewhere [47].

Table 3: Micro-USM process capabilities.

Sr. no.	Features	Capability	Reported by
(1)	Minimum diameter, μm	5	Egashira et al. (1997) [25]
(2)	Maximum (Depth × Length), μm	$60\,\mu$m × $500\,\mu$m	Sun et al. (1996) [22]
(3)	Max. no. of features on single tool	16 microtools of $18\,\mu$m diameter	Egashira et al. (2004) [34]
(4)	Others	Array of pillars ($280\,\mu$m diameter and $6000\,\mu$m depth)	Boy et al. (2010) [42]

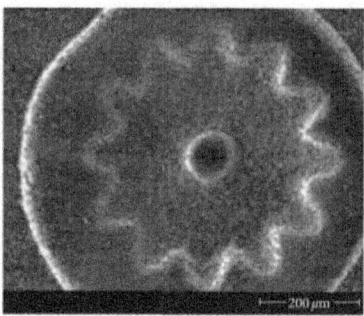

Figure 17: Die-sinking Micro-USMed feature in Alumina [45].

RESEARCH ISSUES

In experimental investigations, micro-USM has been recognized as a promising micromachining process for generating microscale features in (nonconducting) brittle and hard materials. However, research works on micro-USM so far are mainly focused on exploring the feasibility and geometrical ability of machining. One of the possible reasons is that to a certain extent some aspects of micro-USM are similar to macro-USM and can be directly implemented. However, issues with micro-USM need not be similar to that of macro-USM. Few such issues associated with surface finish and removal of debris have already been discussed in Section 4.1. Further, it is quite possible that mechanism of material might get influenced by other factors usually not considered important in macro-USM. Various research issues in the micro-USM process can be categorised as illustrated in Figure 18. Most of these issues are concerning the major task of process improvement for micro-USM.

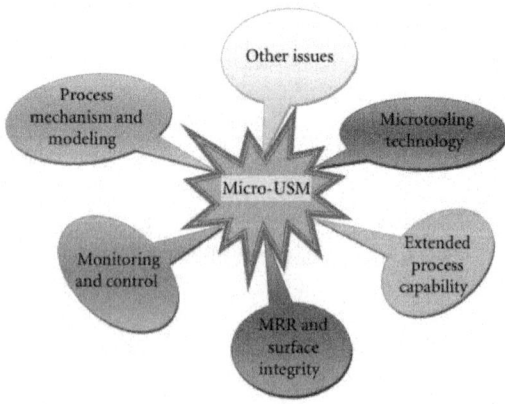

Figure 18: Major research issues in micro-USM.

Interactions between abrasive particles and workpiece are very intricate in micro-USM. A well-structured study for understanding of the material removal mechanism for micro-USM is yet to be carried out. Further analysis of the material removal mechanism associated with the machining process is required. The contribution of each probable phenomenon that might be active during the removal of material including microchipping, abrasion, cavitation, and chemical reaction needs to be investigated. In a microcutting situation, localised temperature might play a significant role too, which call for further research.

The process output parameters (such as material removal rate and surface roughness) of micro-USM depend mainly on the physical/mechanical phenomena at the machining gap. In such processes, many research issues can be originated from the practical requirements (such as surface finish) and existing limitations (such as serious tool wear, particle size, and surface finish) of the process. Studies on process capabilities and process modelling are few aspects that can contribute immensely towards making the process cost effective. Microtooling is another aspect which could contribute significantly towards process capability and such issues need to be addressed early.

Further, although ample process monitoring and control strategies are already in use, the issues might need be relooked in the context of noise reduction and high frequency-low amplitude requirements of the process with adaptive control option owing to finer movements associated.

Other issues can also be summarized as in Figure 19.

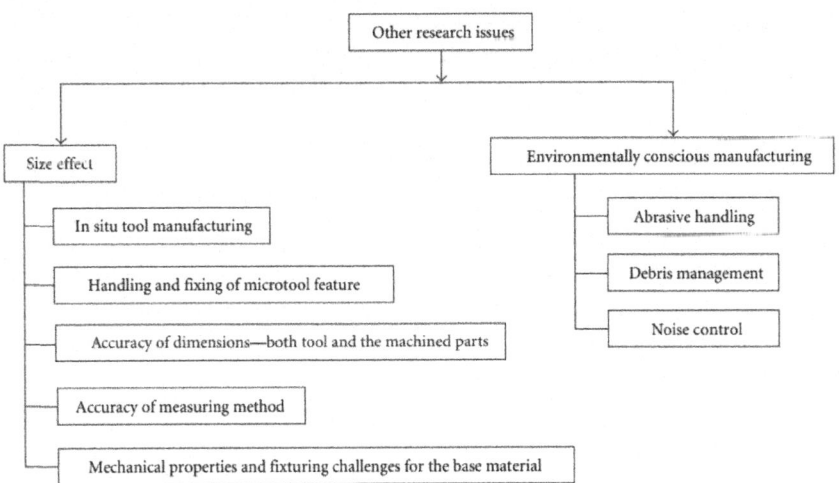

Figure 19: Size effect and environment dominant research paradigms.

Micromachining is not merely reduction in size of the machining conditions such as tool size, abrasive size, vibration amplitude, and feature size. There are other inherent difficulties which should be addressed and treated as research issues. These issues can further be broadly categorised into what is termed as "size effects" and "environmentally conscious manufacturing". There are a number of issues associated with each broad category and can be detailed as illustrated in Figure 19. The size effects include problems like preparation of microtools, their handling, and fixing of the tool feature to the tool holder. The accuracy and performance of micromachining processes depend on the accuracy and appropriateness of the microtool also. However, preparation and physical attachments of such tools are big challenges faced by the micromachining industries. Efficient and accurate methods to develop microtools will significantly influence the processing time, processing quality, and hence the production economy. Focussed researches in this direction can help overcoming such issues. The alignment of the tool plays a vital role and must be accurate to confirm the accuracy of machining. Making a tool with a traditional method (micromilling) and then fixing it with silver brazing or soft soldering to the tool holder does not validate its alignment because the tool is too small for traditional machining and assembling methods. Hence, suitable methods which are capable to fabricate the microtool on the machine itself and maintain the high precision need to be developed. Egashira et al. (1997) have proposed some solutions in the last decade itself and fresh investigations would result further improvements [25].

Another issue included in the size effect is accuracy of dimensions and its measurement methods. Considering the feature size in use (usually, smaller than 500 µm), it is desirable that the measuring equipments are of sufficiently higher. Calibration of such equipment and handling of the micromachined parts for measurements are also challenging. Therefore, in situ microtool preparation and in situ product inspection systems need to be developed in order to minimize the production error and ensure product quality.

The mechanical properties of workpiece material, related handling, and holding difficulties onto the micromachining facility are some other issues that need to be addressed. Multipurpose automatic holding and positioning devices with requisite accuracies need to be developed which will reduce the idle time significantly. The work material used for micromachining is generally very brittle; therefore, proper method of workpiece mounting and controlling of tool is essential to avoid probable damage of the workpiece.

One of the important issues which has hardly been addressed is the environmental aspects. It is one of the requirements of modern manufacturing that the material processing techniques be environment friendly. The noise produced by the ultrasonic frequency generating system itself can be a concern for the human beings in the range. The micro-USM process involves fine abrasives which are certainly not safe for human being to consume. The normally used abrasives (boron carbide, silicon carbide, etc.), although not very well known for their toxicity like the reactants used in the chemical-based processes, can however, cause health hazards during handling. The microchips generated during the process get mixed with the slurry. The efficiency of the process decreases as the debris-mixed slurry is recirculated. Thus, the recirculated slurry contains work material constituents apart from the chemically passive abrasives. Safe handling of this abrasive slurry makes the process more complex. Appropriate methods for separation and/or safe disposal need to be developed to make the process more environment friendly. This debris accumulation in the micromachining zone also results in low MRR and affects dimensional accuracy. Hence, proper debris separation and management method needs to be developed in order to enhance the process efficiency and make the process more environmentally acceptable.

The process capability of micro-USM can further be extended by exploring its application in the area of fabrication of components used in microfluidics. It can be attempted to machine microchannels on glass and silicon for micro-heat exchanger applications and for microsensors. The micro-USM technique has not been so far applied for making of microchannels and can be a topic of research.

POTENTIAL APPLICATIONS

Potential applications of micro-USM with rotated tool include the production of high-aspect-ratio microholes (less than 100 μm in diameter) in silicon and glass wafers, which are in great demand for pressure and flow sensors [43]. Various specific applications include drilling small holes in helicopter power transmission shafts and gears, machining of watch bearings and jewels, slicing semiconductor components, for example, cutting circular wafers and drilling small holes in borosilicate glass. Self-aligned multilayer machining and assembly (SAMMA) is a combined machining and assembly method, which specifies that the machining and the assembly of microparts are fulfilled on the same machine without separate processes. Two 3D microair turbines made up of three layers have been developed using SAMMA and micro-USM [22] and are shown in Figures 20(a) and 20(b). Ultrasonic micromachining can also be efficiently used to micromachine as well as micropolish a tool steel surface with a thermoplastic tool as shown in Figure 21. The micro-USM process can be of extremely useful for the semiconductor industry as the industry needs processing of characteristically brittle materials. The applications of the process can also be explored for processing metal-based materials relatively ductile with suitable adaptations which can be of high demand in microfluidics and heat transfer applications.

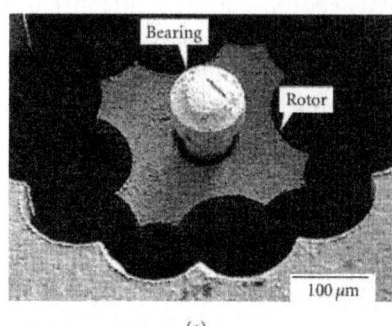

(a)

(b)

Figure 20: (a) Centre-pin bearing airturbine and (b) Micro force-balance airturbine [22].

Figure 21: A micropolished tool steel surface [33].

With the increased awareness of microfluidic physics and the surface science of silicon, polymers, glass, and ceramics, these newer and harder work materials are strongly recommended for the micro-electromechanical system (MEMS). All of these microfluidic systems require the use of micron-sized channels and cavities for liquid or gas delivery and storage. Especially, microchannels and their fabrication technologies are playing an important role in the current development of bio-MEMS. Recently, the present authors have successfully used micro-USM to fabricate such microchannels on glass and silicon as shown in Figures 22(a) and 22(b), respectively. These channels were fabricated using a conventional USM (Sonic Mill) at 20 kHz with silicon carbide abrasives of 20 μm size. Austenitic stainless steel tool heads were used for the trials. The finish of the machined surface will be enhanced considerably with the use of finer abrasives; the currently used 20 μm abrasives appear too coarse.

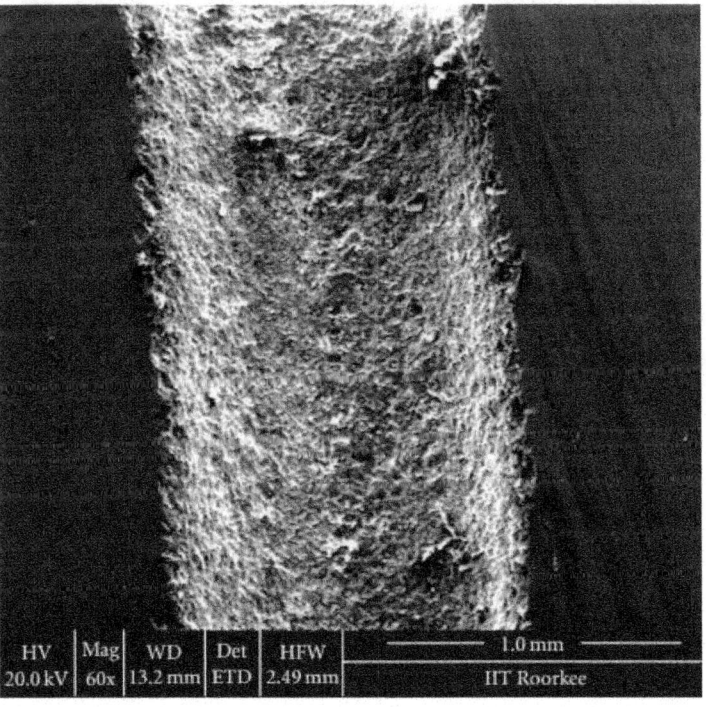

(a)

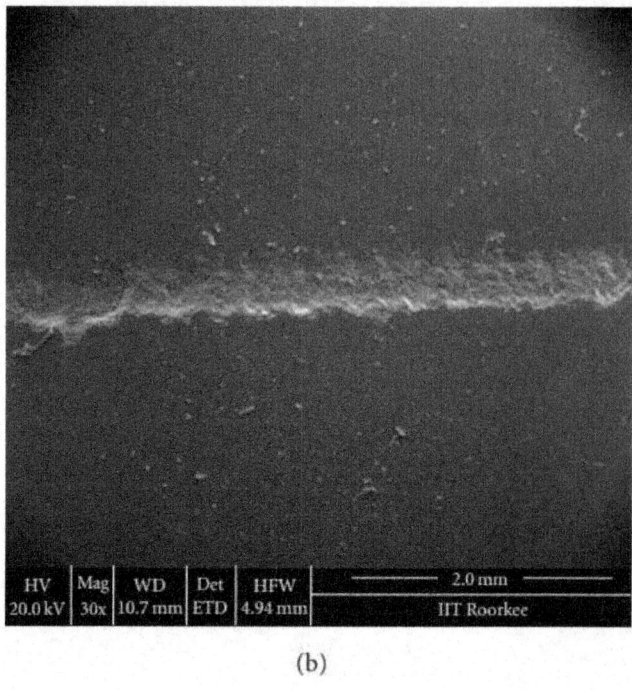

(b)

Figure 22: Microchannels machined using micro-USM on (a) glass and (b) silicon.

CONCLUSION

This paper provides an overview of the main issues concerning different aspects of micro-USM, its performance, and limitations in the application. The paper focuses on the principle of micro-USM, the types of USM processes, tooling, USM process parameters, and the process performance measures, namely, material removal rate and the tool wear ratio. Categorisations of micro-USM processes according to the operating principles used and tool head used have been presented. Mechanics of material removal with reference to the vis-á-vis macro-USM has been illustrated. A comparison with respect to major features of the widely used micromachining processes has been presented. Geometrical capabilities of the micro-USM process have been explored on the basis of reported literature. The presented results can help to plan the process within the expected tolerances. The following major conclusions may be drawn from the present study.

    i.   Microultrasonic machining (micro-USM) is derived from macro-USM, which has been a well-established industrial process offering attractive capabilities to machine hard and brittle materials.

ii.   The contributing mechanisms involved in micro-USM, as in macro-USM, are micro chipping by impact of the free moving abrasive particles, mechanical abrasion by the abrasive particles against the workpiece surface, cavitation effects in liquid agitated by ultrasonic vibration, and chemical actions associated with the liquid being employed.

iii.   The design of tool and transducer capacity plays an important role in providing a resonance state in USM to maximize the material removal rate.

iv.   The hardness of abrasives material should be more than the workpiece as in case of macro-USM. In general, larger abrasive grit sizes and higher slurry concentrations result in higher MRR.

v.   In micro-USM, finer particles might cause deterioration to the surfaces owing to getting pushed into the eroded cavities and inefficient flushing. This is contrary to the general understanding of using finer abrasives for finer surface finish.

vi.   Providing ultrasonic vibrations to the workpiece material while carrying out micro-USM appears more effective contrary to the conventional USM.

REFERENCES

1.   T. Masuzawa, "State of the art of micromachining," CIRP Annals—Manufacturing Technology, vol. 49, no. 2, pp. 473–488, 2000.

2.   E. Gentili, L. Tabaglio, and F. Aggogeri, "Review on micromachining techniques," in Proceedings of the 7th International Conference on Advance Manufacturing Systems and Technology (AMST '05).

3.   J. J. Allen, Micro Electro Mechanical System Design, CRC Press, Boca Raton, Fla, USA; Taylor & Francis, London, UK, 2005.

4.   R. T. Howe and R. S. Muller, "Polycrystalline silicon micromechanical beams," Journal of the Electrochemical Society, vol. 130, no. 6, pp. 1420–1423, 1983.

5.   E. W. Becker, W. Ehrfeld, P. Hagmann, A. Maner, and D. Münchmeyer, "Fabrication of microstructures with high aspect ratios and great structural heights by synchrotron radiation lithography, galvanoforming, and plastic moulding (LIGA process)," Microelectronic Engineering, vol. 4, no. 1, pp. 35–56, 1986.

6.   P. J. French, P. T. J. Gennissen, and P. M. Sarro, "New silicon micromachining techniques for microsystems," Sensors and Actuators, A, vol. 62, no. 1–3, pp. 652–662, 1997.

7.   F. Z. Fang, K. Liu, T. R. Kurfess, and G. C. Lim, "Tool-based micro machining and applications," inMEMS, MEMS/NEMS, Handbook Techniques and Applications, vol. 3, Springer, Berlin, Germany, 2005.

8.   J. L. Liow, "Mechanical micromachining: a sustainable micro-device manufacturing approach?" Journal of Cleaner Production, vol. 17, no. 7, pp. 662–667, 2009.

9.   C. K. Walker, G. Narayanan, H. Knoepfle, et al., "Laser micromachining of silicon: a new technique for fabricating high quality terahertz waveguide components," in Proceedings of the 8th international symposium on Space Terahertz Technology, Harvard University, March 1997.

10. A. A. Tseng, Y. T. Chen, C. L. Chao, K. J. Ma, and T. P. Chen, "Recent developments on microablation of glass materials using excimer lasers," Optics and Lasers in Engineering, vol. 45, no. 10, pp. 975–992, 2007.

11. J. P. Desbiens and P. Masson, "ArF excimer laser micromachining of Pyrex, SiC and PZT for rapid prototyping of MEMS components," Sensors and Actuators, A, vol. 136, no. 2, pp. 554–563, 2007.

12. Z. Y. Yu, K. P. Rajurkar, and H. Shen, "High aspect ratio and complex shaped blind micro holes by micro EDM," CIRP Annals—Manufacturing Technology, vol. 51, no. 1, pp. 359–362, 2002.

13. S. Mahendran, R. Devarajan, T. Nagarajan, and A. Majdi, "A Review of Micro-EDM," in Proceedings of the International Multi Conference of Engineers and Computer Scientists, vol. 2, Hong Kong, March 2010.

14. B. Bhattacharyya, M. Malapati, and J. Munda, "Experimental study on electrochemical micromachining," Journal of Materials Processing Technology, vol. 169, no. 3, pp. 485–492, 2005.

15. R. Singh, M. J. Alberts, and S. N. Melkote, "Characterization and prediction of the heat-affected zone in a laser-assisted mechanical micromachining process," International Journal of Machine Tools and Manufacture, vol. 48, no. 9, pp. 994–1004, 2008.

16. J. C. Hung, B. H. Yan, H. S. Liu, and H. M. Chow, "Micro-hole machining using micro-EDM combined with electropolishing," Journal of Micromechanics and Microengineering, vol. 16, no. 8, pp. 1480–1486, 2006.

17. I. Yang, M. S. Park, and C. N. Chu, "Micro ECM with ultrasonic

vibrations using a semi-cylindrical tool," International Journal of Precision Engineering and Manufacturing, vol. 10, no. 2, pp. 5–10, 2009.

18. T. Endo, T. Tsujimoto, and K. Mitsui, "Study of vibration-assisted micro-EDM-The effect of vibration on machining time and stability of discharge," Precision Engineering, vol. 32, no. 4, pp. 269–277, 2008.

19. S. Koshimizu and I. Iansaki, "Hybrid machining of hard and brittle materials," Journal of Mechanical Working Technology, vol. 17, pp. 333–341, 1988.

20. T. B. Thoe, D. K. Aspinwall, and M. L. H. Wise, "Review on ultrasonic machining," International Journal of Machine Tools and Manufacture, vol. 38, no. 4, pp. 239–255, 1998.

21. D. Kremer, S. M. Saleh, S. R. Ghabrial, and A. Moisan, "The State of the Art of Ultrasonic Machining,"CIRP Annals—Manufacturing Technology, vol. 30, no. 1, pp. 107–110, 1981. · ·

22. X. Q. Sun, T. Masuzawa, and M. Fujino, "Micro ultrasonic machining and self-aligned multilayer machining/assembly technologies for 3D micromachines," in Proceedings of the IEEE Micro Electro Mechanical Systems (MEMS '96), pp. 312–317, 1996.

23. B. Ghahramani and Z. Y. Wang, "Precision ultrasonic machining process: a case study of stress analysis of ceramic (Al_2O_3)," International Journal of Machine Tools and Manufacture, vol. 41, no. 8, pp. 1189–1208, 2001.

24. T. Masuzawa and H. K. Tönshoff, "Three-dimensional micromachining by machine tools," CIRP Annals —Manufacturing Technology, vol. 46, no. 2, pp. 621–628, 1997.

25. K. Egashira, T. Masuzawa, M. Fujino, and X. Q. Sun, "Application of USM to micromachining by on-the-machine tool fabrication," International Journal of Electrical Machining, no. 2, pp. 31–36, 1997.

26. K. Egashira and T. Masuzawa, "Microultrasonic machining by the application of workpiece vibration,"CIRP Annals—Manufacturing Technology, vol. 48, no. 1, pp. 131–134, 1999. ·

27. C. Y. Khoo, E. Hamzah, and I. Sudin, "A review on the rotary ultrasonic machining of advanced ceramics," Jurnal Mekanikal, no. 25, pp. 9–23, 2008.

28. M. Komaraiah and P. N. Reddy, "A study on the influence of work piece properties in ultrasonic machining," International Journal of Machine Tools and Manufacture, vol. 33, no. 3, pp. 495–505, 1993.

29. A. R. Jones and J. B. Hull, "Ultrasonic flow polishing," Ultrasonics, vol. 36, no. 1–5, pp. 97–101, 1998.

30. A. C. Wang, B. H. Yan, X. T. Li, and F. Y. Huang, "Use of micro ultrasonic vibration lapping to enhance the precision of microholes drilled by micro electro-discharge machining," International Journal of Machine Tools and Manufacture, vol. 42, no. 8, pp. 915–923, 2002.

31. R. Singh and J. S. Khamba, "Ultrasonic machining of titanium and its alloys: a review," Journal of Materials Processing Technology, vol. 173, no. 2, pp. 125–135, 2006.

32. T. Kuriyagawa, T. Shirosawa, O. Saito, and K. Syoji, "Micro ultrasonic abrasive machining for three-dimensional milli-structures of hard-brittle materials," in Proceedings of the 16th ASPE Annual Meeting, pp. 525–528, 2001.

33. A. Curodeau, J. Guay, D. Rodrigue, L. Brault, D. Gagné, and L. P. Beaudoin, "Ultrasonic abrasive μ-machining with thermoplastic tooling," International Journal of Machine Tools and Manufacture, vol. 48, no. 14, pp. 1553–1561, 2008.

34. K. Egashira, T. Taniguchi, H. Tsuchiya, and M. Miyazaki, "Microultrasonic machining using multitools," in Proceedings of the 7th International Conference on Progress Machining Technology (ICPMT '04), pp. 297–301, December 2004.

35. Instruction manual for stationary SONIC-MILL 500W Model 2002 (U.S.A).

36. F. B. Gary, Book on Non Traditional Manufacturing Processes, Marcel Dekker, New York, NY, USA, 1987.

37. R. Jadoun, K. Pradeep, and B. K. Mishra, "Taguchi's optimization of process parameters for production accuracy in ultrasonic drilling of engineering ceramics," in Production Engineering, vol. 3, pp. 243–253, German Academic Society for Production Engineering, Springer, Berlin, Germany, 2009.

38. R. S. Jadoun, P. Kumar, B. K. Mishra, and R. C. S. Mehta, "Optimization of process parameters for ultrasonic drilling of advanced engineering ceramics using the Taguchi approach," Engineering Optimization, vol. 38, no. 7, pp. 771–787, 2006.

39. M. Adithan and V. C. Venkatesh, "Study of the performance characteristics of an ultrasonic drilling head," Wear, vol. 33, no. 2, pp. 261–270, 1975.

40. Z. Yu, X. Hu, and K. P. Rajurkar, "Influence of debris accumulation on material removal and surface roughness in micro ultrasonic machining of silicon," CIRP Annals—Manufacturing Technology, vol. 55, no. 1, pp. 201–204, 2006.

41. Z. Y. Yu, K. P. Rajurkar, and A. Tandon, "Study of 3D micro-ultrasonic machining," Journal of Manufacturing Science and Engineering, Transactions of the ASME, vol. 126, no. 4, pp. 727–732, 2004.

42. J. J. Boy, E. Andrey, A. Boulouize, and C. Khan-Malek, "Developments in microultrasonic machining (MUSM) at FEMTO-ST," International Journal of Advanced Manufacturing Technology, vol. 47, no. 1–4, pp. 37–45, 2010.

43. X. Q. Sun, T. Masuzawa, and M. Fujino, "Micro ultrasonic machining and its applications in MEMS,"Sensors and Actuators, A, vol. 57, no. 2, pp. 159–164, 1996.

44. K. Egashira, K. Mizutani, and T. Nagao, "Ultrasonic vibration drilling of microholes in glass," CIRP Annals—Manufacturing Technology, vol. 51, no. 1, pp. 339–342, 2002.

45. C. Zhang, R. Rentsch, and E. Brinksmeier, "Advances in micro ultrasonic assisted lapping of microstructures in hard-brittle materials: a brief review and outlook," International Journal of Machine Tools and Manufacture, vol. 45, no. 7-8, pp. 881–890, 2005.

46. T. Li and Y. B. Gianchandani, "A die-scale micromachining process for bulk PZT and its application to in-plane actuators," in Proceedings of the 18th IEEE International Conference on Micro Electro Mechanical Systems (MEMS '05), pp. 387–390, February 2005.

47. X. Hu, Z. Yu, and K. P. Rajurkar, "State-of-the-art review of micro ultrasonic machining," in Proceedings of the International Conference on Manufacturing Science and Engineering (MSEC '06), October 2006.

Chapter 11

MODULAR MACHINING LINE DESIGN AND RECONFIGURATION: SOME OPTIMIZATION METHODS

S. Belmokhtar, A.I. Bratcu and A. Dolgui

University Campus STeP Ri Slavka Krautzeka 83/A 51000 Rijeka, Croatia

INTRODUCTION

Machining Lines

Automated flow-oriented machining lines are typically encountered in the mechanical industry (Groover, 1987; Hitomi, 1996; Dashchenko, 2003). They are also called transfer (or paced) lines, being preferred mainly for the mass production, as they increase the production rate and minimize the cost of machining parts (Hutchinson, 1976). They consist of a linear sequence of multi-spindle machines (workstations), without buffers in between, arranged along a conveyor belt (transfer system). Each workstation is equipped with several spindle heads, each of these latter being composed of several tools. Each tool executes one or several (for the case of a combined cutting tool) operations. A block of operations is defined by the set of the operations executed simultaneously by one spindle head. When all blocks of a workstation have been accomplished, the workstation cycle time is terminated. The cycle time of the line is the longest workstation cycle time; its inverse is the line's production rate. A machining line designed to produce a single product type is called dedicated line; its optimal structure, once found and implemented, is intended for a long exploitation time and needs high investments. The main drawback of such a system is its rigid structure which does not permit any conversion in case of change in product type. Thus, to react to changes effectively an alternative is to design the system from the outset for all the product types intended to be produced. Research has been conducted to an integrated approach of transfer lines design in the context of flexibility (Zhang et al., 2002). This is the most important aspect which characterizes the potential of a system for reconfiguration (Koren et al., 1999). The chapter deals with the designing of modular reconfigurable transfer lines, where a set of standard spindle heads are used to produce a family of similar products. The objective is to minimise the total investment cost and implicitly minimise the time for reconfiguration. For simplicity, in this chapter, "spindle heads" and "blocks" will have here the same meaning.

Our interest focuses on the configuration/reconfiguration of modular lines. The modularity brought many advantageous: maintenance and overhaul become easier, installation is rapid and reconfiguration becomes possible (Mehrabi et al., 1999). An approach to solve the problem is provided. Such approach is not limited to any specific system. It could be either used to configure a line for one time in case of DML (since the configuration is locked for the whole life time of the system) or to reconfigure the system in case of RMS at each time the demand changes to adapt to the new situation.

The design or configuration of a modular machining lines deals with the selection of modules from a given set and with their assignment to a set of stations. The modules in such lines are the multi-spindle units. Figure 1 illustrates a unit with 2 spindles. When the line has to be configured for the first time, i.e., the line has to be built, the given set of modules is formed on the basis of the following information:

a)   The availability on the market, proposed by the manufacturers of spindle units.

b)   The knowledge of the engineering team to design and manufacture their own spindle units

c)   The already used spindle units which worked on the old lines and are still operational

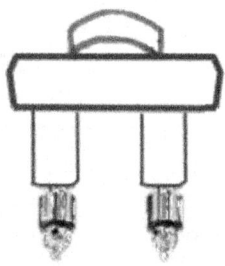

Figure 1: Two-spindle unit.

The problem remains in finding an assignment for spindle units such that all operations are performed and all technological within cycle time constraints are fulfilled. The objective is to minimize the total cost considering the cost of workstations and the costs of spindle units. Depending on the type of system we deal with, the costs can be either the fixed costs in case of dedicated lines or reconfiguration costs considering the amortization of the equipment.

A diagram of the design approach using our IP models is presented in Figure 2. This could be integrated in a holistic approach for line design which is similar to the framework of modular line design suggested by (Zhang et al., 2002).

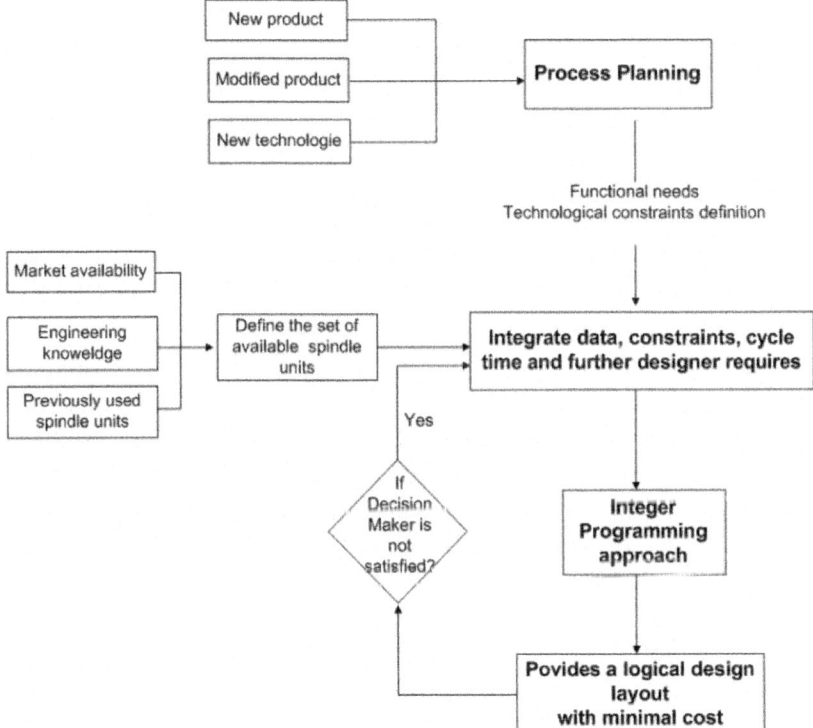

Figure 2: A global conceptual schema.

In the next section related works the problem of designing modular machining lines for single product case is firstly addressed. Then, the proposed approach is generalized to the broader case considering a family of products.

Configuration and Reconfiguration – Related Works

Koren et al. (1999) perform a comprehensive analysis of different types of manufacturing systems. Despite of their high cost, the dedicated machining lines (DML) are very effective as long as the demand exceeds the supply. Even these lines could operate at their full capacity; their average utilisation rate does not exceed 53%, as shown in the cited work. Flexible manufacturing systems (FMS), on the other hand, are built with the maximal available flexibility. They are able to respond to market changes, but are very expensive due to the involved CNC technology. The same study shows that 66% of FMS are not exploiting their full flexibility. Consequently, capital lies idle and an important portion is wasted.

A new alternative to the latter systems is brought by the reconfigurable manufacturing systems (RMS). The RMS aims to compensate the disadvantages of the last systems. This can be achieved by combining the high productivity of DML and flexibility of FMS, hence, providing a cost-effective and quick response to market changes. The cited authors define the RMS as being "designed at the outset for rapid change in structure, as well as in hardware and software components, in order to quickly adjust production capacity and functionality within a part family in response to sudden changes in market or in regulatory requirements."

Youssef & ElMaraghy (2005) identify two aspects of the reconfiguration, namely the software part and the hardware (physical) part. The effort is placed in the first part – machine re-programming, re-planning, re-scheduling, rerouting – whereas the physical reconfiguration relies upon adding/removing machines and changing the handling material. Son (2000) proposes a genetic algorithm based design methodology of an economic reconfiguration in response to demand variations, by using a configuration similarity index. RMS must necessarily be based on a modular structure to meet the requirements for changeability. To configure machining systems the interfaces between the modules have prior importance, therefore these latter must meet some standard specifications. A first step to reconfigurability and, meanwhile, its strongest justification, is to ensure the possibility of producing a family of products, instead of a single one, such that to enable a smooth re-adaptation of the system to a continuously changing demand. A low cost design of a mixedmodel machining line will implicitly ensure the re-adaptation time minimisation also.

A single-part versus multiple-part manufacturing systems (SPMS vs. MSPS) critical analysis is performed by Tang et al. (2005a). The SPMS concern the production of a single type of product on a practically rigid line configuration, whereas the MSPS are intended for a product family. The MSPS are obviously more complex, need a more sophisticated transfer system, but conversely offer more flexibility at a lower cost. In the following, we give a formal description of the designing machining line problem for single product.

Single Product Case

Problem Description

In order to model the problem we have to understand the mechanism of the machining process. We consider the lines where the activation of the spindle units at workstations is sequential. At the level of a workstation, the spindle units operate one after another on the positioned part to be manufactured. So, each workstation has an execution time equal to the sum total of its spindle times. The cycle time of the line is the elapsed time between the starting machining of the spindle units and their end on all workstations. Thus, the cycle time is determined by the slowest station of the line. At the end of each cycle time, the parts are moved to the next station and another cycle begins. Figure 3 illustrates such a line with 2 workstations. The first workstation is equipped with 2 multi-spindles whereas the second has only one unit. Each unit is composed of two spindles.

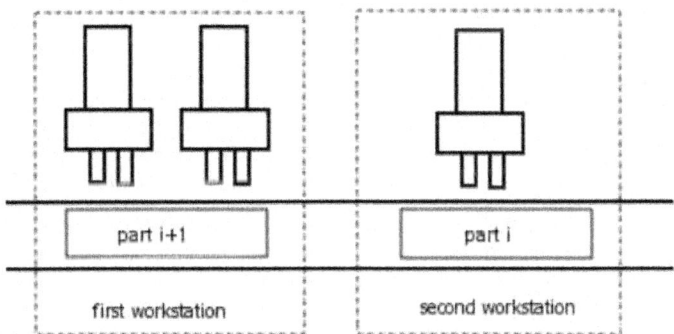

Figure 3: An example of a dedicated line.

Notations and assumptions are as follows:

-   N is the set of operations that have to be performed on each part (drilling, milling, boring, etc.).

-   $B = \{B_r | B_r \subset N\}$ corresponds to the set of all available multi-spindle units, where each is defined by the subset of operations it performs. A multispindle unit B_r is physically composed of one or several tools performing simultaneously corresponding operations. For the sake of simplicity, the term block is used henceforth to refer to a multi-spindle unit. Thus, the block $B_r \subset N$ is said to contain the operations performed by the corresponding multi-spindle unit.

- q_r is the cost of block B_r,

- t_r is the execution time of block B_r,

- C_s is the average cost for any created workstation,

- C_T is the cycle time to not exceed,

- m_0 and n_0 are the maximum number of workstations which can be created and the maximum number of blocks which can be assigned to any workstation, respectively.

It is assumed that the following constraints are known:

1. cycle time,

2. precedence relations between operations,

3. exclusion conditions for blocks and

4. inclusion constraints for operations to be executed at the same workstation.

The above constraints are defined as follows:

1. An upper bound on the cycle time insures a minimal threshold of throughput.

2. Precedence relations impose an order which should be respected between some operations. For example, before drilling or boring one part a jig boring should be performed. A jig boring consists in making a notch when "true position" locating is required. The order relation over the set N can be represented by an acyclic digraph $G^{or} = (N, D^{or})$. An arc $(i, j) \in N \times N$ belongs to the set D^{or} if the operation j must be executed after operation i.

3. Exclusion conditions correspond to the incompatibility between some operations, e.g. it can be the inability to activate some tools on the same workstation. The same kind of constraints have been already studied by Park, Park and Kim (1996) where the assignment of tasks may be restricted by some incompatibilities (minimum or maximum distances in terms of time or space between stations performing a pair of tasks). In our case, this incompatibility is extended to blocks such

that blocks involving incompatible operations are not assigned to the same workstation. The constraints are represented by a collection D^{bs} of subsets $D_l \subseteq B$ such that all blocks from the set D_l cannot be assigned to the same workstation. But any subset strictly included in D_l can be assigned to the same workstation.

4. Restrictions related to operations which have to be executed on the same station are referred to as inclusion relations. For example, if a precise distance is required between two holes, the operations corresponding to their drilling should be performed at the same workstation. If these operations are performed on different workstations, then the impact of moving reduces greatly the chance of successful precision drilling for subsequent holes. The inclusion conditions can be represented by a family D^{os} of subsets $D_t \subseteq N$ such that all operations of the same subset D_t from D^{os} must be performed at the same workstation. In Pastor and Corominas (2000) similar restrictions are considered, these operations are introduced as one operation. Beyond the possibility of merging the operations, we also consider the case where operations can be performed separately with different spindle units (if such units are available).

The Integer Linear Program

Decision variables are defined as follows:

$$x_{rk} = \begin{cases} 1, & \text{if block } B_r \text{ is assigned to station } k \\ 0, & \text{otherwise} \end{cases}$$

$$y_k = \begin{cases} 1, & \text{if station } k \text{ is opened} \\ 0, & \text{otherwise} \end{cases}$$

Additional parameters have to be defined, they are described as follows.

- $K(r) = [head_r, tail_r] \subseteq [1, m_0]$ is the interval of the workstation indices where block $B_r \in B$ can be assigned. The $head_r$ is the earliest station where block B_r can be assigned and tailr is the last; $head_r$ and $tail_r$ values are computed on the basis of problem constraints. Obviously, the number of decision variables is directly proportional to the width of the interval $K(r)$;

- $Q(i) = \{B_r \in B \mid i \in B_r\}$. Thus, $Q(i)$ contains all blocks from B which perform operation $i \in N$;

- interval KO(j) corresponds to all stations where operation j can be performed:

$$KO(j) = \bigcup_{B_r \in Q(j)} K(r)$$

The objective function is expressed as follows:

$$\text{Minimize} \sum_{k=m'+1}^{m_0} C_s \cdot y_k + \sum_{k=1}^{m_0} \sum_{B_r \in B} q_r \cdot x_{rk}$$

The following constraints ensure for each operation from set N its execution in only one workstation:

$$\sum_{B_r \in Q(i)} \sum_{k \in K(r)} x_{rk} = 1, \quad \forall i \in N$$

The cycle time constraints for each workstation are:

$$\sum_{B_r \in Q(i)} \sum_{k \in K(r)} x_{rk} = 1, \quad \forall i \in N$$

The precedence constraints must not be violated:

$$\sum_{B_r \in Q(i)} \sum_{l=head_r}^{k-1} x_{rl} \geq \sum_{B_r \in Q(j)} x_{rk}, \quad \forall (i,j) \in D^{or}, \quad \forall k \in KO(j)$$

The inclusion constraints for operations are respected with the following constraints:

$$\sum_{B_r \in Q(i)} x_{rk} = \sum_{B_r \in Q(j)} x_{rk}, \quad \forall \{i,j\} \subseteq D_l, \quad \forall D_l \in D^{os}, \quad \forall k \in KO(j)$$

The exclusion constraints for blocks are respected if:

$$\sum_{B_r \in D_l} x_{rk} \leq |D_l| - 1, \quad \forall D_l \in D^{bs}, \quad \forall k \in \bigcap_{B_r \in D_l} K(r)$$

The maximal number of blocks by workstation is respected by:

$$\sum_{B_r \in \{B_s \in B | k \in K(s)\}} x_{rk} \leq n_0, \quad \forall k = 1, 2, \ldots, m_0$$

The following constraints are added in order to avoid the existence of intermediate empty workstations:

$$y_{k-1} - y_k \geq 0, \quad k = m^* + 2,....,m_0$$
$$y_k \geq x_{rk}, \quad k \geq m^* + 1, \quad B_r \in \left\{ B_s \in \mathbf{B} \middle| k \in K(s) \right\}$$

In the next section we show how the model for a single product, above presented, can be extended to a family of products. Many of the assumptions and notation are maintained, and some new assumptions have to be considered, as shown below.

Family Product Case

Problem Description

The features of the product family are supposed known, each product being described by the corresponding precedence graph and the required cycle time. An admissible resemblance degree must be assumed between products, as to some well known rules of defining a product family (Kamrani & Logendran, 1998) (for example, they must have a minimal number of common operations). The goal is to design the minimal cost line configuration from the given set of available modules. This configuration must ensure a desired throughput level. By designing, we mean to determine the number of workstation to establish and to equip them with blocks such that all operations are executed only once. We are interested in the best structure of such line with regard to fixed cost point of view. Thus, the objective function is a linear combination of the cost of workstations to be established and the costs of blocks chosen to be assigned to them.

Problem Assumptions

For our problem the following assumptions are adopted for the whole family:

- a set of all possible blocks is known and it is authorized that each block may be used only partially; the operations from a block are executed in parallel;

- each operation must be executed only once, by a single block assigned to exactly one workstation;

- the activation of blocks belonging to the same station is sequential;

- station setting cost, blocks' operating times and cost of each block are given.

An admissible assignment of blocks to stations is to find such that all the technological constraints – order, compatibility and desired cycle times – be satisfied and the total equipment cost (for all stations and blocks) is minimal. We are not interested to find the blocks activation sequence for each product, but we should ensure for the provided solution that such an order always exists for each product.

Related Literature

This kind of problems is known in literature as Line Balancing problems (Scholl & Klein, 1998). In case of a single product type and if each block is composed of only one operation, then the problem is reduced to the basic problem, Simple Assembly Line Balancing (SALB). The aim is to minimize the unbalance (cost) of the line for a given line cycle time. The unbalance is minimal if and only if the number of stations is minimal.

In general, integer linear programming models for the SALBP are formulated and solved by exact or heuristic methods (Baybars, 1986; Talbot et al., 1986; Johnson, 1988; Erel & Sarin, 1998; Rekiek et al., 2002). The Mixed-model Assembly Line Balancing (MALB) problem approaches the optimization of lines with several "versions" of a commodity in an intermixed sequence (Scholl, 1999). The design of such lines – also called multi-product or multi-part lines – must take into account the possible differences between versions – among others, different precedence relations, different task times, etc. By enriching the basic assumptions of SALB, the Generalized ALB (GALB) problems have also been stated, in order to solve more realistic problems – a comprehensive survey may be found in Becker & Scholl (2006).

The balancing problems with equipment selection have some common features with the studied problem. In this context, a recent approach proposes the use of a genetic algorithm for configuring a multi-part optimal line, having the maximal efficiency (minimal ratio of cost to throughput) as criterion for the fitness function (Tang et al., 2005b). Our problem differs essentially from the balancing and equipment selection problems because the operations are simultaneous into blocks. This feature makes it impossible to directly solve by the known methods.

The closest problems are studied in Dolgui et al. (1999), Dolgui et al. (2000), Dolgui et al. (2001), Dolgui et al. (2005) and Dolgui et al. (2006b) where all blocks at the same station are executed sequentially (block by block) and any alternative variants of blocks are not given beforehand (any subset of the given operation set is a potential block). In Dolgui et al. (2004),

Belmokhtar et al. (2004), Dolgui et al. (2006a) the blocks are known and are executed in parallel. All these papers concern the case of a single product, for which three solving approaches have been proposed: a constrained shortest path; mixed integer programming (MIP); heuristics. A generalization of the linear programming approach to the case of a product family and sequential activation of blocks was for the first time presented by Bratcu et al. (2005).

The rest of this chapter is organised as follows. In Section 2 a detailed formal description of the problem is presented, along with the needed notations and explanations on how the constraints' aggregation is made. In Section 3 the proposed solving procedure is discussed, based upon a linear programming model, possibly to improve by some reductions. Section 4 is dedicated to some concluding remarks and to perspectives.

FORMAL STATEMENT OF THE PROBLEM

Input Data and Notations

The problem is identified by answering to the following questions:

- what must be produced? – this is the set of features characterizing the product family (number of products, set of operations and precedence constraints for each product);

- how should them be produced? – these are the blocks' characteristics (cost and operating time of each block);

- production conditions (external environment – like demand, for example – and internal constraints – for example, maximal number of stations or maximal number of stations on the line).

As consequence, the following input data are given for each instance of the problem:

- - p is the number of product types to manufacture;

    - N_i is the set of operations corresponding to product i, i=1,2,...p;

    - $N = \bigcup_{i=1}^{p} N_i$ is set of operations of the whole family;

- - B is the set of blocks for realizing the operations from N, with R being the set of B's indices;

    - Cbr is the cost of block r and tbr is its operating time;

- - Cs0 is the cost of setting a new station;

    - m0 is the maximal number of workstations and n0 is the maximal num ber of blocks assigned to a station;

- Δt is the time interval in which the demand of product i is ni, i=1,2,…p.

From quantitative information about the demand, that is, from Δt and n_1, the imposed cycle times for each product, T_{ci}, may be computed. It is assumed that B contains only blocks having operating times smaller than the smallest cycle time of the products: $tb_r \le \min_{i=1,2,…p} \{Tc_i\}$ for all r ∈ R.

Three types of technological constraints are considered:

- the precedence constraints
- the inclusion constraints
- the exclusion constraints.

Their meanings and formalizations are detailed hereafter:

- The precedence relation is a partial order relation over each set N_i. It is represented by an acyclic digraph $G_i=(N_i,Dor_i)$. One should notice that the precedence relation is here taken in the non strict sense: a vertex $(j,k) \in N_i \times N_i$ belongs to the set Dor_i if either operation k must be executed after operation j, or the two operations are performed in parallel (in the same time).

- Exclusion conditions correspond to incompatibility between some operations and have the same meaning like in the single product case (see section 1.3.1).

- Restrictions related to operations which have to be executed on the same station are referred to as inclusion relations. These also have the same meaning like in the single product case.

The inclusion conditions can be represented by a family D_i^{os} of subsets $D_t \subseteq N_i$ such that all operations of the same subset D_t from os D_i should be performed at the same workstation. One can note that each D_i^{os} is at most a partition over the set N_i. Remark: In the general case, where the blocks of operations are not known, there exist inclusion and exclusion constraints of assigning operations to the same block, that is, sets of operations forbidden to be assigned to a block all together, respectively sets of operations which are mandatory to be assigned to a same block. For our problem, these constraints are taken into account while forming the block set B, therefore it is supposed that all the blocks of B already meet these constraints.

Aggregated Constraints

As all the products should be produced on the same line, using the same equipment, an initial phase in dealing with a reconfigurable line optimization is the aggregation of the constraints concerning the individual products. Due to the assumption on the same characteristics for products belonging to the same family, the constraints should not be contradictory. In case where it happens, there will be no feasible solution to the design problem: a line for machining the given product family cannot be designed under the given constraints.

In particular, there should normally not be contradictory precedence constraints between operations common to several products of the family. But if this however happens, one must first make the aggregation of the precedence constraints to obtain a single precedence graph for the whole family. This operation will influence also the other two types of constraints, as shown later. The aggregated (or total) precedence graph is obtained by merging together the sets of individual precedence relations, according to the following steps:

- represent all graphs superposed and merge the multiple vertices in the same sense;

- delete redundant arc (i,j), i.e., if there is a path from i to j (containing several transitive vertices), then the arc (i,j) is said to be redundant and consequently should be deleted;

- identify the circuits due to contrary precedence between operations in different individual graphs; the nodes from these circuits correspond to operations that cannot be separated without violating the precedence constraints, therefore, such operations are merged together into the newly introduced macro-operations;

- redraw the total acyclic graph, where the macro-operations are represented by ordinary nodes.

- The definition of the macro-operations will consequently induce changes in all the operation sets, N_i, i=1,2...p, and also in the total set, N, as well as in the set of both inclusion and exclusion constraints.

Concerning the inclusion constraints, each element of each D_i^{os}, i=1,2,...p, containing only some operations of a macro-operation, is extended with the absent operations. These sets are then united and the elements having non empty intersection are merged together. Furthermore, the blocks which execute just a part of the macro-operations should be eliminated; the final, aggregated set of inclusion constraints is denoted by Dos. Next, the sets of exclusion constraints (elements of Dbs) containing these blocks are to be eliminated too. These two latter actions may also be viewed as part of the model reduction (detailed in Section 3.2).

Example

Here below is detailed an example to illustrate the aggregated constraints.

Let a product family be composed of p=3 products, given by their precedence graphs, as in Figure 4. The corresponding sets of operations are:

N_1={1,2,3,4,5,6,7,8,9,10,11}, $|N_1|$=11,
N_2={1,2,3,4,5,6,7,8,9,10,11,12}, $|N_2|$=12,
N_3={1,2,4,5,7,8,9,10,11,12,13}, $|N_3|$=11.

The total operation set is therefore:

$$N = \bigcup_{i=1}^{p} N_i = \{1,2,3,4,5,6,7,8,9,10,11,12,13\}, \ |N|=13,$$

and there are 9 common operations for all products:

$$N_b = \bigcap_{i=1}^{p} N_i = \{1,2,4,5,7,8,9,10,11\}$$

Ellipses in Figure 4 represent the inclusion constraints (defining which operations must be performed on the same workstation) for product i, D_i^{os}. The corresponding sets are:

$$D_1^{os} = \left\{ \underbrace{\{7,8\}}_{D_{11}^{os}}, \underbrace{\{9,10\}}_{D_{12}^{os}} \right\}' \quad D_2^{os} = \left\{ \underbrace{\{7,8\}}_{D_{21}^{os}}, \underbrace{\{9,10,12\}}_{D_{22}^{os}} \right\} \quad \text{and} \quad D_3^{os} = \left\{ \underbrace{\{7,8\}}_{D_{31}^{os}}, \underbrace{\{9,10\}}_{D_{32}^{os}} \right\};$$

to these ones a supplementary set of inclusion constraints is added, $D_s^{os} = \{\{3,13\}\}$, concerning two operations belonging to different products, which have to be together when the products are realized on the same line, This latter set is suggested by dotted rectangle in Figure 4. Taking into account the aggregation of constraints, this set will be treated just like any other D_i^{os}.

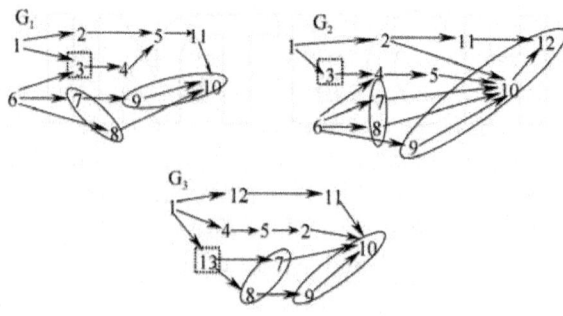

Figure 4: Precedence graphs and inclusion constraints for a family of 3 products.

Next, suppose that the total set N is to be executed by a set B of 12 multifunctional tools (blocks), whose features – operations to perform, operating times and costs – are provided in Table 1 (abbreviations t.u. and m.u. denote respectively time units and monetary units).

Table 1: Set of blocks to manufacture the product family described in Figure 4

Block r	Operations	Operating time, tb_r [t.u.]	Cost, Cb_r [m.u.]
B_1	{1,3,6,13}	9	250
B_2	{1,3,13}	8	170
B_3	{1,2,7,8}	6	281
B_4	{2,5,9}	10	150
B_5	{2,5,7,8,11}	9	275
B_6	{2,6,9,10}	11	230
B_7	{4,6,8,10}	13	211
B_8	{4,7,8}	9	160
B_9	{4,7,8,9}	10	215
B_{10}	{5,12,13}	6	158
B_{11}	{10,11,12,13}	12	230
B_{12}	{2,5,10,11,12}	11	260

The fixed cost of setting a new station is C_{s0}=350 m.u., the maximal number of stations is m_0=5 and the maximal number of blocks per station is n_0=3.

Next, suppose that the following constraints related to minimal line throughput are imposed: in a period of Δt=48300 t.u. n_1=2100 pieces of the first product , n_2=1932 pieces of the second product and n_3=2300 pieces of the third product must be manufactured. Computing the required cycle with the expression $T_{ci}=\Delta t/n_i$ for time for product i, one obtains T_{c1}=23 t.u., T_{c2}=25 t.u. and T_{c3}=21 t.u. respectively.

The exclusion constraints are provided by a unique block set for all products. Suppose that the sets of blocks forbidden to be assigned to the same station are:

$$D^{bs} = \left\{ \underbrace{\{B_1, B_6, B_7, B_{10}\}}_{D_1^{bs}}, \underbrace{\{B_1, B_9\}}_{D_2^{bs}}, \underbrace{\{B_3, B_5, B_{11}\}}_{D_3^{bs}} \right\}$$

The constraints aggregation is part of a pre-processing performed on the initial data about the problem; it will lead to a single set of each type of constraints, distinguished by the exponent "pp" (acronym corresponding to preprocessing). The generation of the total precedence graph – a single one for the whole product family – by merging together the individual precedence graphs, is the starting point in aggregating constraints. In the considered case, this operation allows to identifying two circuits, $2 \rightarrow 5 \rightarrow 2$ and $11 \rightarrow 10 \rightarrow 12 \rightarrow 11$, due to contrary precedence relations between same operations in different individual graphs. Therefore, two macro-operations are formed, denoted by a={2,5} and b={10,11,12}. The total precedence graph, G, defining a partial order relation on the new set of operations, N^{pp}={1,3,4,a,b,5,6,7,8,13}, is shown in Figure 5.

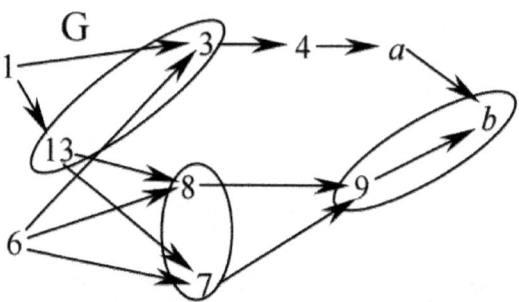

Figure 5: Precedence graph of the whole product family.

As the macro-operations have been introduced, the operation set for each product has also changed:

$$N^{pp1}=\{1,a,3,4,6,7,8,9,b\}, \ |N^{pp1}|=9,$$

$$N^{pp2}=\{1,a,3,4,6,7,8,9,b\}, \ |N^{pp2}|=9,$$

$$N^{pp3}=\{1,a,4,7,8,9,b,13\}, \ |N^{pp3}|=8.$$

Also, those blocks which cannot execute but only some operations from a macro-operation must be eliminated. These are B_3, B_5, B_6, B_7 and B_{10}. Therefore, the new set of blocks is $B^{pp}=\{B_1,B_2,B_4,B_8,B_9,B_{11},B_{12}\}$, having noted that now B4={a,9}, B11={b,13} and B_{12}={a,b}.

Next, one must perform the aggregation of the inclusion constraints, taking into account the existence of macro-operations. Thus, each element of each D_i^{os}, i=1,2,3, and each element of D_s^{os} containing only some of the operations included in macro-operations is first extended with the absent operations:

$$D_1^{os \, pp} = \left\{ \underbrace{\{7,8\}}_{D_{11}^{os \, pp}}, \underbrace{\left\{9, \underbrace{10,11,12}_{b}\right\}}_{D_{12}^{os \, pp}} \right\}$$

$$D_2^{os\,PP} = \left\{ \underbrace{\{7,8\}}_{D_{21}^{os\,PP}}, \underbrace{\left\{ 9,10,11,12 \atop \underbrace{}_{b} \right\}}_{D_{22}^{os\,PP}} \right\}$$

$$D_3^{os\,PP} = \left\{ \underbrace{\{7,8\}}_{D_{31}^{os\,PP}}, \underbrace{\left\{ 9,10,11,12 \atop \underbrace{}_{b} \right\}}_{D_{32}^{os\,PP}} \right\}$$

Set $\underbrace{D_j^{os\,PP} = \{3,13\}}_{D_n^{os\,PP}}$ remains unchanged. Then, all these 4 sets are united and the elements having non empty intersection are merged together. The aggregated set of inclusion constraints is:

$$D^{os\,PP} = \left\{ \underbrace{\{3,13\}}_{D_1^{os\,PP}}, \underbrace{\{7,8\}}_{D_2^{os\,PP}}, \underbrace{\{9,b\}}_{D_3^{os\,PP}} \right\}$$

The set of exclusion constraints, Dbs, must be changed because of the elimination of some blocks in the previous aggregation steps. Each element of Dbs has the meaning of forbidding the blocks to be all together on the same station (note that the global exclusion relation does not necessarily mean mutually exclusion). Hence, if a block happens to be eliminated, then all the exclusion constraints containing it will also be eliminated. In the considered example, sets D_1^{bs} and D_3^{bs} are those to be eliminated. Therefore, the final exclusion constraints set is:

$$D^{bs\,PP} = \left\{ \underbrace{\{B_1, B_9\}}_{D_1^{bs\,PP}} \right\}$$

SOLVING BY INTEGER LINEAR PROGRAMMING (IP)

IP Formulation

The cost optimization of a reconfigurable machining line admits a IP formulation. The presented model is an extension of the one built for the single product case (Dolgui et al., 2004; Belmokhtar et al., 2004). The main difference is that individual constraints are aggregated for all products, the blocks work sequentially in each station and the precedence relation is not strict (see above). The model needs that the following variables and additional parameters be introduced:

- binary decision variables x_{rk}, with $x_{rk}=1$ if block r is assigned to station k and $x_{rk}=0$ otherwise, $k=1,\ldots, m_0$;
- $y \geq 0$ to denote the number of stations;
- m* to denote a lower bound of the number of stations;
- for each block r, the interval $K(r)=[\text{head}(r),\text{tail}(r)]$, with head(r) being the earliest station and tail(r) being the latest station where block r can be assigned;
- family $F_s=\{F_1,\ldots F_v\}$ of pairs of blocks having common operations: $F_q=\{r,t\}$ such that $B_r \cap B_t \neq \emptyset$ for any $q \in V=\{1,\ldots,v\}$ – i.e., only one block from each pair of F_s can be used in a decision; F_s is called the subset of (pairs of) alternative blocks;
- $F_0 = B \setminus \bigcup_{q \in V} F_q$ – i.e., F0 is the set of blocks that will surely appear in the solution;
- $w_{rt}=|B_r \cap D_t|$ and $W_t=\{r \in \mathbf{R} | w_{rt}>0\}$ for any block r and any $D_t \in Dos$, that is, W_t are the blocks able to execute the operations belonging to subset D_t of aggregated inclusion constraints;
- $U_t=\{r \in \mathbf{R} | i_t \in B_r\}$, where it is a given operation from the set $D_t \in Dos$;
- for each block B_r, the set M(r) of operations not belonging to B_r which directly precede the operations of B_r;
- for each block r, the set $H(r)=\{t \in \mathbf{R} | B_t \cap M(r) \neq \emptyset\}$, containing the blocks capable of performing the operations from M(r);
- $H=\{r \in \mathbf{R} | M(r) \neq \emptyset\}$, i.e., the set of operations having predecessors;
- $htr=|Bt \cap M(r)|$ for any $r \in H$ and any $t \in H(r)$;
- R* to denote an upper bound of the set of blocks to be assigned to the last station of the line.

The objective function corresponds to the line total investment cost minimization:

$$Cs_0 \cdot y + \sum_{r=1}^{|\mathbf{R}|} \sum_{k=1}^{m_0} Cb_r \cdot x_{rk} \rightarrow \min \tag{1}$$

which, for reasons of speeding up computation, can be also expressed as:

$$Cs_0 \cdot y + \sum_{r=1}^{|R|} \sum_{k=1}^{m_0} (Cb_r + \varepsilon_r k) x_{rk} \to \min$$

(2)

where ε_r is a sufficiently small nonnegative value. The optimization is subject to a set of constraints, whose mathematical forms are given and explained hereafter.

The first constraints ensures the execution of every operation from the aggregate operation set, N, in exactly one station. Both cases are considered: either choosing blocks without intersection with the others (from F_0), or choosing alternative blocks (from elements F_q of F_s):

$$\sum_{k \in K(r)} x_{rk} \leq 1, \quad r \in F_0$$

(2)

$$\sum_{r \in F_q} \sum_{k \in K(r)} x_{rk} \leq 1, \quad q \in V$$

(3)

As all the operations from the total set N must be executed, it holds that:

$$\sum_{r \in R} \sum_{k \in K(r)} |B_r \cap N| \cdot x_{rk} = |N|$$

(4)

The aggregate precedence constraints on set N impose that:

$$\sum_{t \in H(r)} \sum_{s \in K(t), s \leq k} h_{tr} \cdot x_{ts} \geq |M(r)| \cdot x_{rk}, \quad r \in H, \ k \in K(r)$$

(5)

The aggregate inclusion constraints for the stations are met if:

$$\sum_{r \in W_t} w_{rt} \cdot x_{rk} = |D_t| \cdot \sum_{s \in U_t} x_{sk}, \quad D_t \in Dos, \ k \in \bigcup_{s \in U_t} K(s)$$

(6)

Respect of the aggregate exclusion constraints for assigning blocks to the same station writes as:

$$\sum_{r \in D_t} x_{rk} \leq |D_t| - 1, \quad D_t \in Dbs, \ k \in \bigcap_{s \in D_t} K(s)$$

(7)

As n_0 is the maximal number of blocks to be allocated to a workstation, then:

$$\sum_{r \in \{t \in R | k \in K(t)\}} x_{rk} \leq n_0, \quad k = 1, 2 \ldots m_0$$

(8)

The constraints concerning the number of stations require that:

$$y \geq k \cdot x_{rk}, \quad r \in R^*, \quad k \in K(r), \quad k \geq m^* \tag{9}$$

The last constraints impose that the cycle time requirements be met:

$$\sum_{r \in \{t \in R | k \in K(t)\}} t_r \cdot x_{rk} \leq Tc_i, \quad i=1,2,...p, \quad k \geq m^* \tag{10}$$

In the above model one can note the dependence of the number of stations, y, on the variables x_{rk}. The model does not explicitly claim the integrality constraint on y, but constraint (9) and the objective function (1) implicitly force it. Some possible model reductions may be performed, in order to minimize the number of decision variables, as proposed below.

Reduction of Model and Computation of Bounds

In order to reduce computation time, an analysis of the block set after performing the aggregation of constraints – that is, after identifying the macrooperations – can allow some supplementary block eliminations. The steps presented hereafter are not mandatory, but can contribute to avoid useless computation.

The first action is to check if situations like the one described in Figure 6a) happen. In this figure, the precedence relations between two operations from different blocks are such that a "block circuit" appears. Obviously, a solution cannot contain all the blocks involved in such a circuit, but it is however sufficient that a single block be deleted. It is proposed that a heuristic elimination rule be used in this case, namely the most expensive – as cost per operation – block be eliminated, which is consistent with the goal of the total investment cost minimization. Note that such eliminations must start from the maximal circuits identified.

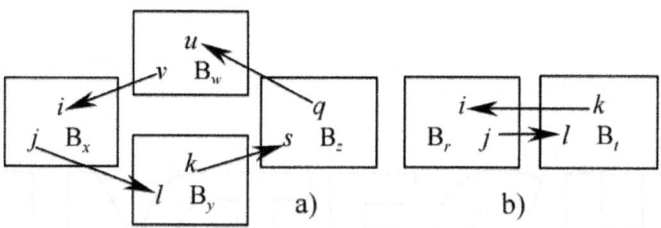

Figure 6: Example of blocks forming circuits and loops.

A special case is that of two vertices circuits (loops). If two blocks r and t form a loop (like in Figure 6b)), it is not necessary that one of them be deleted, but certainly only one will appear in a solution. It is therefore sufficient to treat them as alternative blocks (i.e., the pair (r,t) be an element of F_s).

The second step of the model reduction concerns also the set of blocks. Remember that this set, B, will have already undertaken some changes due to the constraints aggregation, as above mentioned.

Thus, for each block Br from the last block set B, a subset B' is searched, such that:

- operations from B' give the total set, N;
- $|B'| \leq m_0 \cdot n_0$;
- all blocks from B' are mutually disjoint.

Each block for which such a subset, B', does not exist must be eliminated. Even if these reductions are not performed before the optimization phase, the optimizer will implicitly make them. But in large scale problems this could negatively affect the computation time. Hereafter are presented the reductions possible for the example considered in Section 2.3.

The first reduction step is to check the existence of "circuits" on subsets of B^{pp}. It is said that a precedence relation exists between two blocks if and only if all the operations from a block precede all the operations from the other block. In Figure 7 precedence relations between two blocks have been represented by thick arrows, whereas the thin arrows denote precedence relation between operations.

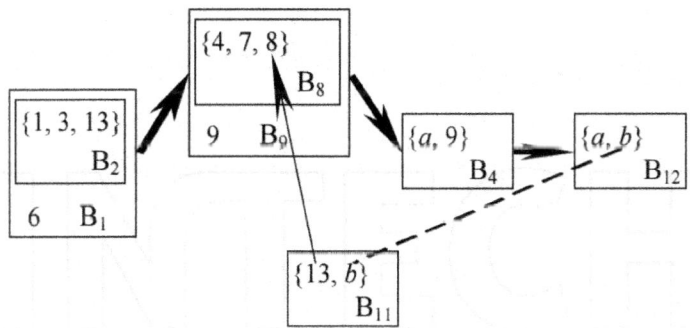

Figure 7: Circuits on the set of blocks after the constraints aggregation, B^{pp}.

One can remark the existence of a circuit of blocks, that is, $B_8 \rightarrow B_4 \rightarrow B_{12} \rightarrow B_{11} \rightarrow B_8$. According to the heuristic of eliminating the most expensive block as cost per operation, block B_{11} (57.5 m.u. per operation) must be eliminated, as to data provided in Table 1, Hence, the reduced block set is: $Bpp = \{B_1, B_2, B_4, B_8, B_9, B_{12}\}$, which still contains all the operations of N^{pp} (if this were not be the case, then the problem would not have any solution).

Concerning the second reduction step, this is not very important in small scale problems. But for large scale problems, it may be useful to make it in the preprocessing phase. In the analyzed case, one can verify that blocks B_2, B_4 and B_8 may be eliminated.

As for the computation of bounds, we briefly present here below how the intervals K(r) are computed for any block r, as well as the minimal number of stations, m*, and the maximal block set, R*, to be assigned to the last station. Intervals K(r) and m* are computed based upon the algorithm proposed by Dolgui et al. (2000), using the notion of distance between two operations. In the general case, this distance takes one of three values (0, 1 or 2) – in our case, it can take only two values: either 2, if the two operations can only be performed by blocks forbidden to be on the same station (i.e., belonging to elements of Dbs), or 0, otherwise.

In the cited work, the blocks are not a priori known. Therefore, the problem is solved in two steps: first determining the bounds of assigning operations to blocks, then for allocating blocks to workstations.

Thus, the algorithm begins with computing the values $q^-(i)$ and $q^+(i)$ for any operation i, which denote the earliest and respectively the latest block where operation i can be assigned. In our case, to compute values $q-(i)$, the algorithm needs as input data the total precedence graph, G, the aggregated inclusion constraints, Dos, and the distance matrix, d ($|N|\times|N|$). Values $q+(i)$ result from the same algorithm, but entering the reversed precedence graph, G^r. In the second step, values $k^-(i)$ and $k^+(i)$ of the earliest and the latest station where operation i can be assigned are computed, using the relation:

$$k^{+/-}(i) = [q^{+/-}(i)/n_0],$$

with [·] denoting the smallest integer larger or equal with the argument. For any block r, there are finally computed:

$$\begin{cases} head(r) = \max\{k^-(i)|\ i \in B_r\} \\ tail(r) = \min\{k^+(i)|\ i \in B_r\} \end{cases}$$
(11)

The lower bound on the number of stations results as:

$$m^* = \max\{k^-(i)|\ i \in \mathbf{N}\}$$
(12)

Having computed intervals K(r) for any block r, a sufficiently good value of R* may result from the following algorithm.

- tail_max ← max{tail(r)| $B_r \in B$}
- Find the minimum head of the blocks having the tail equal to tail_max. Let be head_min this minimum.

- Form the subset of blocks having the tail strictly larger than head_min. This subset is R*.

An immediate goal aimed at in the near future is to improve the value of R*.

CONCLUSION AND PERSPECTIVES

This chapter has approached the problem of optimizing the investment cost of modular machining lines (also called transfer lines) aimed at producing a family of products. The possibility of allowing variations over the set of products is the most important step for a manufacturing system to become reconfigurable. The specificity of the lines analyzed here is the parallelization of the operations' execution by the same spindle head. Due to the important investment cost required to build such lines, the search of an optimal design decision for the whole family appears as necessary. The potential economic benefits achieved are not negligible and is one of the motivating reasons to propose such an approach. The powerful mathematical programming tools make it possible to solve exactly and efficiently such problem, providing cost effective solutions. However, searching for the optimal solution may be prohibitively time-consuming, as much as the scale problem is larger.

The cost optimization of this kind of machining lines is a new and poorly studied problem, different from the classical SALB problem, but also NP-hard. This work presented a complete mathematical formulation of the problem as a linear program and proposed a procedure to follow for obtaining an exact (optimal) solution. An important phase of the solving procedure is the aggregation of constraints, which practically allows that the studied problem be treated like the single product one. Some proposals of model reduction have also been presented, to avoid running time exhaustion. A particular attention should be focused on improving the bounds. Due to the exponential complexity of the integer linear programming solving algorithm, a bad behavior when increasing the problem's dimension is highly possible. The large number of constraints is a feature that will potentially allow the coupling of the presented exact method with different types of heuristics, able to provide good bounds to exact methods. We consider that, for applying the proposed method in real life environments, this coupling is definitely necessary.

REFERENCES

1. Baybars, I. (1986). A survey of exact algorithms for the simple line balancing problem, Management Science, Vol. 32, pp. 909-932, ISSN 0025-1909

2.   Becker, C. & Scholl, A. (2006). A survey on problems and methods in generalized assembly line balancing, European Journal of Operational Research, Vol. 168, No. 3, (1 February 2006), pp. 694-715, ISSN 0377-2217

3.   Belmokhtar, S.; Dolgui, A.; Guschinsky, N.; Ihnatsenka, I. & Levin, G. (2004). Optimization of transfer line by constraint programming approach, Proceedings of Computer and Industrial Engineering Conference (CD-ROM), November 13-16 2004, San Francisco, U.S.A.

4.   Bratcu, A. I.; Dolgui, A. & Belmokhtar, S. (2005). Reconfigurable Transfer Lines Cost Optimization – A Linear Programming Approach, Proceedings of the 10th IEEE International Conference on Emerging Technologies and Factory Automation – ETFA 2005, Lo Bello, L. & Sauter, T. (Eds.), pp. 625-632, ISBN 0-7803-9402-X, Catania, Italy, September 19-22 2005, IEEE, Piscataway, NJ, U.S.A.

5.   Dashchenko A. I. (Ed) (2003). Manufacturing Technologies for Machines of the Future 21st Century Technologies, Springer.

6.   Dolgui, A.; Guschinsky, N. & Levin, G. (1999). On problem of optimal design of transfer lines with parallel and sequential operations, Proceedings of the 7th International Conference on Emerging Technologies and Factory Automation – ETFA'99, J. M. Fuertes (Ed.), pp. 329-334, ISBN 0-7803- 5670-5, Barcelona, Spain, October 18-20 1999, IEEE, Piscataway, NJ, U.S.A.

7.   Dolgui, A.; Guschinsky, N. & Levin, G. (2000). Approaches for transfer lines balancing, Preprint No. 8, Institute of Engineering Cybernetics/ University of Technology of Troyes

8.   Dolgui, A.; Guschinsky, N.; Levin, G. & Harrath, Y. (2001b). Optimal design of a class of transfer lines with blocks of parallel operations, Proceedings of the IFAC Symposium on Manufacturing, Modeling, Management and Control MIM 2000, P. P. Groumpos, A. P. Tzes (Eds.), pp. 36-41, ISBN 0080435548, Patras, Greece, July 12-14, 2000, Elsevier.

9.   Dolgui, A.; Guschinsky, N.; Levin, G.; Louly, M. & Belmokhtar, S. (2004). Balancing of Transfer Lines with Simultaneously Activated Spindles, Preprints of the IFAC Symposium on Information Control Problems in Manufacturing – INCOM'04 (CD-ROM), April 5-7 2004, Salvador da Bahia, Brazil (to appear also in the IFAC Proceedings Volume, Elsevier)

10.   Dolgui, A.; Finel, B.; Guschinski, N.; Levin, G. & Vernadat, F. (2005). A heuristic approach for transfer lines balancing. Journal of Intelligent Manufacturing, Vol. 16, No 2, pp. 159-171, ISSN 0956-5515

11. Dolgui, A.; Guschinsky, N. & Levin, G. (2006a). A special case of transfer lines balancing by graph approach. European Journal of Operational Research Vol. 168, No. 3, (1 February 2006), pp. 732-746, ISSN 0377-2217

12. Dolgui, A.; Finel, B.; Guschinski, N.; Levin, G. & Vernadat, F. (2006b). MIP Approach to Balancing Transfer Lines with Blocks of Parallel Operations, IIE Transactions, 2006, (In Press)

13. Erel, E. & Sarin, C. (1998). A survey of the assembly line balancing procedures, Production Planning and Control, Vol. 9, pp. 414-434, ISSN 0953-7287

14. Groover, M. P. (1987). Automation, production systems and computer integrated manufacturing, Prentice Hall, Englewood Cliffs, New Jersey

15. Hitomi, K. (1996). Manufacturing system engineering, Taylor & Francis

16. Hutchinson, G. (1976). Production Capacity: CAM vs. transfer line, IE, September 1976, pp. 30-35

17. Johnson, J. R. (1988). Optimally balancing large assembly lines with FABLE, Management Science, Vol. 34, pp. 240-253, ISSN 0025-1909

18. Kamrani, A. K. & Logendran, R. (1998). Group technology and cellular manufacturing: methodologies and applications, Gordon and Breach

19. Koren, Y.; Heisel, U.; Jovane, F.; Moriwaki, T.; Pritschow, G.; Van Brussel, H. & Ulsoy, G. (1999). Reconfigurable Manufacturing Systems. CIRP Annals, Vol. 48, pp. 527-598, ISSN 0007-8506

20. Park, K.; Park, S. & Kim, W. (1996). A heuristic for an assembly line balancing problem with incompatibility, range and partial precedence constraints. Computers and Industrial Engineering, Vol. 32, No 2, pp. 321-332, ISSN 0360-8352

21. Pastor, R. & Corominas, A. (2000). Assembly line balancing with incompatibilities and bounded workstation loads. Ricerca Operativa, Vol. 30, pp. 23-45, ISSN 0390-8127

22. Rekiek, B.; Dolgui, A.; Dechambre, A. & Bratcu, A. (2002). State of art of assembly lines design optimization, Annual Reviews in Control, Vol. 26, No. 2, pp. 163-174, ISSN 1367-5788

23. Scholl, A. & Klein, R. (1998). Balancing assembly lines effectively: a computational comparison. European Journal of Operational Research, Vol. 114, pp. 51-60, ISSN 0377-2217

24. Scholl, A. (1999). Balancing and sequencing assembly lines, 2nd edition, Physica, Heidelberg

25. Son, S. Y., (2000). Design Principles and Methodologies for Reconfigurable Machining Systems, Ph.D. Dissertation, University of Michigan

26. Talbot, F. B.; Paterson, J. H. & Gehrlein, W. V. (1986). A comparative evaluation of heuristic line balancing techniques, Management Science, Vol. 32, pp. 430-454

27. Tang, L.; Yip-Hoi, D. M.; Wang, W. & Koren, Y. (2005a). Selection Principles on Manufacturing System for Part Family, Proceedings of the 2005 CIRP International Conference on Reconfigurable Manufacturing (CD-ROM)

28. Tang, L.; Yip-Hoi, D.; Wang, W. and Koren, Y. (2004). Concurrent linebalancing, equipment selection and throughput analysis for multi-part optimal line design. The International Journal for Manufacturing Science & Production, Vol. 6, Nos. 1-2, pp.71-81, ISSN 0793-6648

29. Youssef, A. Y. M. & ElMaraghy, H. A. (2005). A New Approach for RMS Configuration Selection, Proceedings of the 2005 CIRP International Conference on Reconfigurable Manufacturing (CD-ROM)

30. Zhang, G. W.; Zhang S. C. & Xu, Y. S. (2002). Research on flexible transfer line schematic design using hierarchical process planning. Journal of Materials Processing Technology, Vol. 129, pp. 629-633, ISSN 0924-0136

Chapter 10

A MANUFACTURING FRAMEWORK FOR CAPABILITY-BASED PRODUCT-SERVICE SYSTEMS DESIGN

Gokula Vijaykumar Annamalai Vasantha[1] , Hitoshi Komoto[2] , Romana Hussain[3] , Rajkumar Roy[4], Tetsuo Tomiyama[4] , Steve Evans[5] , Ashutosh Tiwari[4] and Stewart Williams[6a]

[1]Design, Manufacture and Engineering Management, University of Strathclyde, Glasgow G1 1XQ, UK

[2]Advanced Manufacturing Research Institute, National Institute of Advanced Industrial Science and Technology, Tsukuba, Ibaraki 305-8564, Japan

[3]School of Engineering, Cranfield University, Cranfield, Bedfordshire MK43 0AL, UK

[4]Manufacturing and Materials Department, Cranfield University, Cranfield, Bedfordshire MK43 0AL, UK.

[5]Institute for Manufacturing, University of Cambridge, Cambridge CB3 0FS, UK

[6]Welding Engineering and Laser Processing Centre, Cranfield University, Cranfield, Bedfordshire MK43 0AL, UK

ABSTRACT

Manufacturers aim to design product-service systems (PSS) which integrate services with products to attain sustained competitive advantage from a life cycle perspective. PSS design should be customised solutions which are aligned to integrated stakeholders' capabilities, a subject which the extant literature has not sufficiently addressed. This paper proposes a systematic framework for the PSS solution provider to address this aim and operationalizes this through software developed for PSS design which models stakeholders' individual activities and simulates their occurrences depending on their relations. The framework stresses that integrated stakeholders' capabilities define continuing ability to generate a desired operational outcome for the customers. The paper reports a PSS design case for a laser system manufacturer and then applies the framework to it. The industrial experts' views on this framework reveal that it helps to develop PSS design from a holistic systems approach which facilitates a change in the designer's mindset from a product-centric to a systems-centric.

The level of trust and transparency required for this framework is argued to be absent in most industrial sectors, being one of the foremost limitations for implementation of PSS.

BACKGROUND

Particularly in the light of recent economic downturns, manufacturers require alternative strategies to cope with globalization and reduced profit margins, and for retaining and attracting customers. One promising approach in helping manufacturers to achieve these objectives is product-service systems (PSS). This approach facilitates manufacturers in bundling products and services together to create a sustained competitive advantage. It aims at providing more value by fostering the optimal use of resources which can be sustained for both consumption and production. Major advantages for the PSS solution provider include prolonged and strategic relationships with the customer and product/ service improvements based on the improved understanding of customer needs. However, Neely's [1] findings from the analyses of a large industrial database were that designing, implementing and managing PSS is a huge challenge to the provider as there are distinct possibilities of economic downturns.

A review of current PSS literature [2] reveals that the theories and methodologies to aid the design of PSS are still in their initial stages of development, and substantial research is required to develop a practical PSS design methodology along with supporting tools. Also, from the interviews which were conducted with 15 industrial maintenance experts of large technical systems such as aerospace engines and related systems, naval ships, land vehicle systems, trains and trucks, our understanding is that, currently, PSS conceptual design in practice is *ad hoc* (fairly intuitive) and lacks a systematic approach in allowing heterogeneous (tangible and intangible characteristics) aspects to be reflected within the PSS design process. The interaction of the stakeholders in the design process and the unique characteristics of products, services, networks of players as well as the supporting infrastructure [3] which are involved in the design of PSS all demand new theories, methodologies, tools and techniques.

The aim of our research is the formal development of a PSS design framework for the PSS solution provider to create customised PSS designs aligned to integrated stakeholders› capabilities. The proposed PSS framework has a view to developing PSS designs to support long-term business solutions from a capability viewpoint whilst stressing the development of additional value to customers by fostering the optimal use of integrated stakeholders› resources. A framework has been developed to address gaps identified in the literature. The framework encompasses a systems thinking perspective, which aims to improve overall PSS design on a system level and avoid sub-optimized solution towards any single activity across the whole life cycle-including remanufacturing. Therefore, this work did not focus specifically on the remanufacturing phase. To help PSS designers› model and simulate PSS designs, a framework has been implemented in a software environment. Service CAD integrated with a life cycle simulator (ISCL) [4] has been chosen because it provides both modelling and simulation facilities. A simulation facility is used to quantitatively evaluate the performance of the PSS designs regarding the requirements. This paper is structured into seven sections detailing: the gaps identified in the literature, understanding industrial challenges in PSS development, defining the constituents of PSS design, a step-by-step illustration of the proposed framework, application of the proposed framework to an industrial laser system case study, academic and industrial experts› views on this framework, and conclusions with future directions of research. Figure 1 illustrates the structure of the paper with the research methodology description.

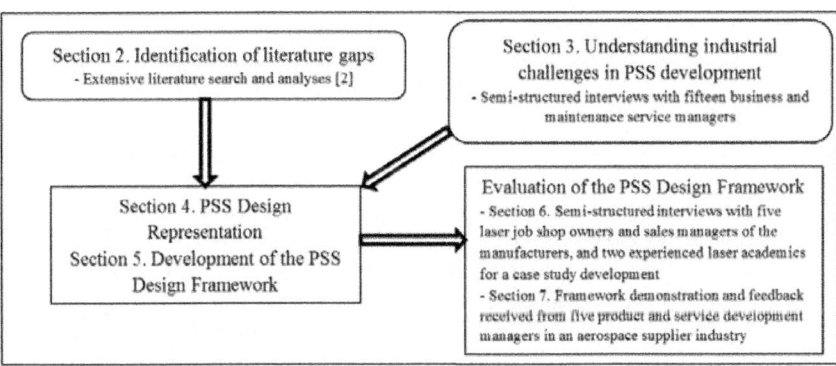

Figure 1: Structure of the paper with the research methodology description.

RELATED PSS DESIGN RESEARCH

In this paper, four state-of-the-art methodologies which have been proposed for PSS design are reviewed: Komoto and Tomiyama [4, 5] proposed a method to design and analyze business models of manufacturers by focusing on the variations of services in product life cycles (e.g. rental, sharing, maintenance and upgrade services, and pay-per-function sales). The business models are modelled using Service CAD and are quantitatively analyzed using a life cycle simulation technique [6]. Sakao and Shimomura [7] and Shimomura et al. [8, 9] developed Service Explorer for service engineering to design products with a higher added value from enhanced services. Service Explorer is also a Service CAD software which employs discrete event simulation methods for the evaluation of PSS design. The scope of evaluation is different from the sequence of activities of customers in a specific service environment (e.g. restaurants) [9] to those of the activities of manufacturers and users of products during their entire life cycle [4]. Maussang et al. [10] presented a PSS design model to assist engineers in the joint development of physical products and interacting services to generate more added value. Alonso-Rasgado et al. [11] and Alonso-Rasgado and Thompson [12] proposed a total care design process to develop innovative offerings consisting of hardware and services integrated to provide complete functional performance. Compared to other methodologies in the literature, these four methodologies are detailed, demonstrated through industrial examples, published in refereed journals and widely discussed in the literature. Table 1 compares four state-of-the-art PSS design methodologies with reference to six characteristics. The chosen six characteristics are important for discussion based on which the proposed framework is developed and nurtured. Table 1 illustrates the differences between the approaches proposed in the literature and aids to identify the literature gaps.

Table 1: Comparison of state-of-the-art PSS design methodologies

Characteristics of the proposed methods	Komoto and Tomiyama [4,5]	Shimomura et al. [7-9]	Maussang et al. [10]	Alonso-Rasgado et al. [11,12]
PSS definition	A set of services in the life cycle of products, whose characteristics are customized with respect to the services	Service/product engineering as a discipline seeking to increase the value of artefacts by focusing on service	PSS are composed of physical objects and service units that relate to each other	Total care products as integrated systems comprising hardware and support services
Aim	Support the design and analysis of integration of services with a product life cycle and the identification of the characteristics of products	Focuses on service engineering to design products with a higher added value from enhanced services	Assists engineers in the joint development of physical products and interacting services to generate more added value	Develop innovative offerings consisting of hardware and services integrated to provide complete functional performance
The first step	Define goal(s) and quality as specified by product users	Define the state change of the receiver	Customer expectations, needs and specifications involved in the whole life cycle	Business ambitions of the client
PSS variables	Stakeholders in a product life cycle and the activities (e.g. production, use and services)	RSP, sequential chain of agents, relationships among RSPs, function, entity and attribute parameters	External functional representation, specifications of the physical elements	Customer's business needs, business solutions, clearer view of the hardware and/or services
PSS design representation technique	A graph description based on service formulation	Business process markup language, service blueprint	Scenarios and FBD	No representation technique is mentioned
Evaluation of PSS designs	Life cycle simulation considering multi-objectives (e.g. economic and environmental)	AHP, Dematel and Petri nets (discrete event simulation)	No evaluation approach proposed	Business case validation and evaluation of alternatives

RSP, receiver state parameter; FBD, functional block diagram; AHP, analytic hierarchy process.

This table shows that all of the authors define PSS in terms of increasing the value of hardware (functional entities that carry out the elementary functions of the system) by focusing on services (entities that will ensure the smooth functioning of the whole system). Some of the aforementioned methods can deal with capabilities of stakeholders as parameters of PSS models employed in their methods. However, they do not provide specific methods or guidelines for PSS design based on the measurement, control and increase of capabilities. Furthermore, a common problem in PSS definition is the usage of different terms to define constituents of PSS. Various terms such as environment, activity, provider, receiver, channel, content, receiver state parameter (RSP), agent, and relationship among RSPs, function, entity and attribute parameters can create confusion and misunderstanding with regards to defining and communicating PSS design amongst research and industrial practitioners - a simple and unified PSS definition illustrating its constituents is required. Also, a PSS design representation technique should be commonly accepted to implement and develop a computer-supported PSS design platform for the effective evaluation of PSS performance and the capture and reuse of PSS design knowledge. The next section summarizes the industrial challenges in the PSS development.

Industrial challenges in designing PSS

This section summarizes the challenges which have been observed by the experts in business and maintenance of several providers of large, technical, capital-intensive and sensored product-service systems. Specifically, these are challenges observed in maintenance planning. The maintenance experts were interviewed to understand engineering services issues because maintenance service occupy nearly three quarters of acquisition and support costs as compared to other aftermarket services [13]. Semi-structured interviews were held with five companies: aerospace engines and related systems, naval ships, land vehicle systems as well as, trains and trucks. Overall, manufacturers generally lack the competence to address the challenges of PSS-type contracts. The reasons for this situation are listed below:

- Customers now have higher expectations from manufacturers.
- Product-orientated manufacturers tend to be product-centric and so do not have the mindset to develop and deliver PSS.
- There tends to be very few technical employees of manufacturers interacting with customers.
- Customer issues could take months or years to be resolved if they are to be addressed by services.
- Current PSS design methodologies tend to be *ad hoc* and tend not to start with the business case.
- The design of the product and service is not completely performed simultaneously and maintenance is mostly an afterthought; only slight modifications to the product are considered following a decision to create a PSS.
- Manufacturers tend not to perform enough modelling to fully understand maintenance activities sufficiently to undertake PSS-type contracts.

- There is a lack of high-level strategic decisions to, for example, trade-off between design, maintenance and supply network solutions for the efficiency of the overall solution.
- A common understanding of PSS-type contracts is lacking across teams.
- The framing of competitive maintenance offerings is a challenge. Most importantly, the consideration of value-added benefits to customers and a suitable operative model between stakeholders to ensure the throughput of inputs as well as reasonable profits tends to be lacking.

These challenges present obstacles to the design of PSS solutions offered by these companies. As a result of an investigation of the state-of-the-art methodologies and challenges in designing PSS in the literature, we have framed the following research questions to be answered in this paper:

- What constitutes PSS design? (This question intends to define characteristics and properties of the system).
- How can customised PSS solutions be designed to be aligned to integrated stakeholders› capabilities?

The next section discusses our definition and constituents of PSS design.

WHAT CONSTITUTES PSS DESIGN?

PSS design aligned to integrated stakeholders' capabilities is mandatory to achieve a viable and sustained solution for an intended duration. A capability can be defined as the continuing ability to generate a desired operational outcome [14]. The definition of capability exemplifies how the joint capability of all the stakeholders could achieve the desired outcome required by the PSS customer and for the PSS provider to design economically sustainable PSS. Considering capabilities as a core element, we defined PSS design as a process to synthesize and create sustained functional behaviour through tangible products and intangible services. Sustained functional behaviour represents the degree to which a system can continuously achieve its purpose by adapting its capabilities. To represent sustained functional behaviour, an activity-based modelling approach is proposed (Figure 2). An activity could be defined as an action incorporated or influenced in the customer's system. The activity-based modelling approach is in-line with the definition of the PSS design processes provided by Komoto and Tomiyama [4], in which designers define the activity to meet a specified goal and quality, and also define environment as being the circumstance within which that activity is realized. Furthermore, Matzen [15] and Tan [16] emphasized activity systems in modelling PSS development. Tan focused on customer activity cycles whereas Matzen viewed activity systems with a broader and general view encompassing both customer and company activities. Both Matzen and Tan conceptualized PSS solutions by considering artefact-, activity- and actor-based domains with slight differences. In this work, capability is mapped through resources, competences, responsibilities undertaken and outcomes. Inputs to a particular activity are mapped through customer needs and precedence activities' outcomes. Other influential parameters on an activity are enforced through an environmental variable. In Figure 2, the coloured boxes represent modification incorporated in the existing system.

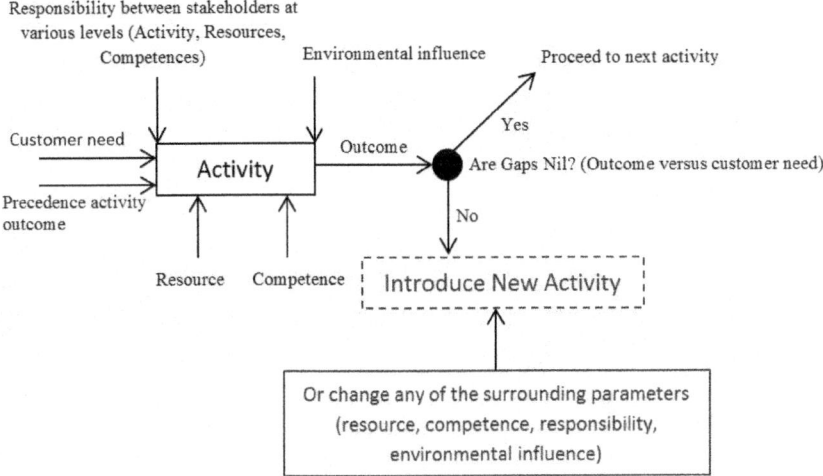

Figure 2: Representation of PSS design constituents (coloured boxes represent modification to the existing system).

We propose in this work that a network of seven parameters, namely *activities, customer needs, environmental influences, resources,competences, responsibilities, inputs (precedence activities' outcomes) and outcomes*, are sufficient to model PSS design. We have chosen these familiar and simpler terminologies for easy acceptance within industrial practitioners. These parameters map the required information for mapping products, services, processes and interactions of all the elements involved within the system. Mapping of these parameters helps to understand the gaps within the current customer's system (PSS user) and aids the development of innovative PSS designs to satisfy customer business needs. The derivation of these parameters in PSS design is presented in the proposed framework. The framework is detailed step by step in the next section.

A CAPABILITY-BASED PSS DESIGN FRAMEWORK

The proposed framework aims to support manufacturers in designing customer-adjusted PSS designs which are aligned to stakeholders' capabilities. The core principles supporting this framework to realize this aim are as follows:

- Gap analysis: Identification of value addition required in the customer's system (PSS user) by understanding their needs through assessing customer's business activities and constructing relative key performance indicators [17]. This initial step lays a foundation to develop customised solutions.

- Generation of new and/or re-designs of integrated product and service solutions along with conditions and consequences of each design. These designs take into consideration the partial substitution of product and service shares over the life cycle.

- Responsibility assignment which considers the capabilities of all the stakeholders involved at various levels: activity, object (resources), and parameter (competences). This assignment enables the derivation of innovative function-, availability- or result-oriented business models.

- Synthesis-generated solutions in each gap to improve the overall PSS design on a system level and avoid a sub-optimized solution towards any of single activity, stressing the importance of resource effectiveness.

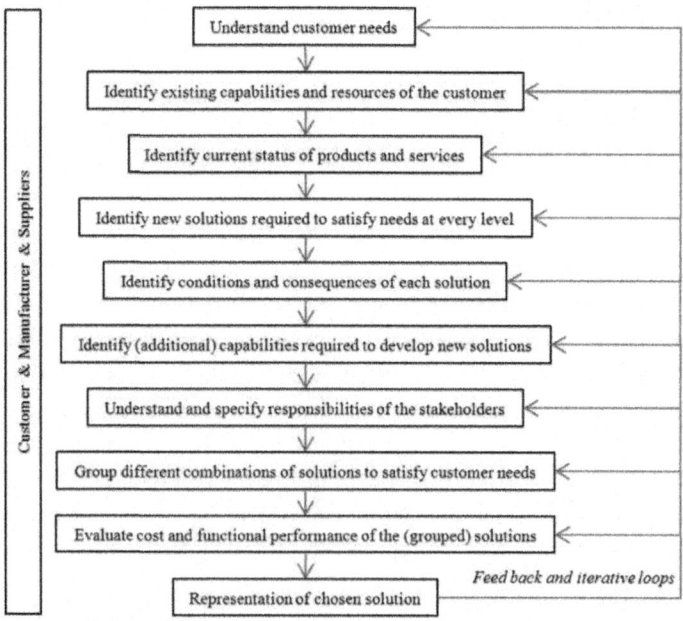

Figure 3: The proposed capability-based PSS design framework.

- Standard representation of PSS designs is required for effective communication across all the stakeholders.

These core principles are structured into 10 steps in the proposed framework (Figure 3). The steps mentioned in this framework are highly inter-dependent, and feedback loops exist between every step. The following sub-sections describe each step individually and detail the course of action involved.

Step 1: understand customer needs

The first step in deriving customer needs from their complete business activities and existing systems is not taken into consideration in most of the PSS methodologies; commonly, customer needs are deemed to be the requirements of products and services. Note that the term 'customer' represents the PSS user throughout this framework. Figure 4 illustrates a typical product-centric life cycle. The product life cycle within the customer's business process only partly covers operation. Although the consideration of the product life cycle improves the understanding of what is required of products and services, extra value could be offered by considering the customer's goals as revealed by their business processes (Figure 4). Moreover, understanding the customer's business processes reveals the 'need behind the need' [18] of the customer that has to be fulfilled. Tan [16] has emphasized that PSS solutions may be conceptualized by considering the product life phase, customer activities and actor network. Figure 4 illustrates the point that it is the consideration of the capabilities that are required to use a product that presents opportunities to add value. Identifying and understanding the customer's overall needs should therefore be the foremost step in the design process. Once the overall customer needs have been identified by focusing on their business processes, the next step would be to identify the current capabilities of the customer.

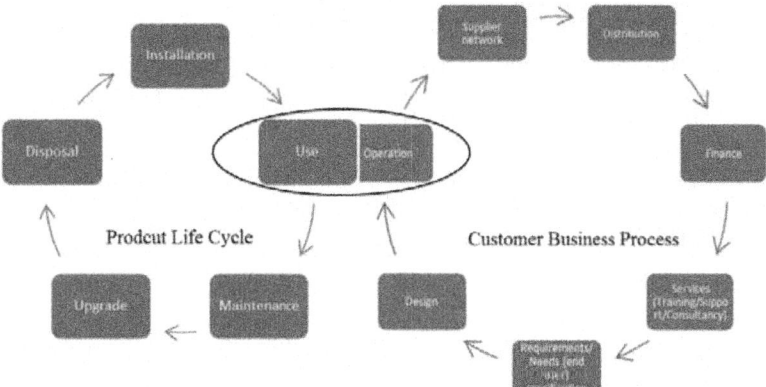

Figure 4: Integration of product life cycle in customer business process.

Step 2: identify existing capabilities and resources of the customer

Identifying the customer's needs is followed by understanding of the customer's existing capabilities. Such an understanding helps to develop designs which are more aligned to their capabilities. A capability can be defined as the continuing ability to generate a desired operational outcome [14]. Capabilities can be realized through people, processes, tools and technology [19]. It should be noted that these elements are highly coupled and should be visualized together. Integration of these elements will be facilitated if the customer's activities are identified and the efficiency of each activity is measured. For this reason, capabilities should be mapped depicting their outcomes and the reasons for deficiency. Some of the parameters to assess each activity could be performance, time taken, reliability, responsiveness, expense and quality [20, 21]. Such analyses will highlight the gaps within the customer capabilities that need to be filled by a PSS design. In the next step, in-depth analyses of existing products and services are performed to ascertain the degree to which the needs can be addressed.

Step 3: identify current status of products and services

The identification of existing products and services (whether they are on the market or just being developed) that could help to address the identified customer needs is mandatory. The next step is to identify and highlight how these offerings could be changed or added to meet customer needs even more closely; this would involve identifying sub-systems within those products and services that need to be considered to achieve this change. This outcome could be achieved by:

- Comparing the key performance indicators for these products and services against those of the solution to meet customer needs.
- Performing root cause analyses to find out which sub-system within the products or services is responsible for failing to fully meet customer needs.

The data required for this step is available in most organizations through condition based monitoring and effective data management systems. The outcomes from these points should help to inform design as to how these offerings could be improved. It should help in the design of the right product and service mix to satisfy the needs of customers. For example, Figure 5 illustrates the capability shifts between machine capability and maintenance service. In scenario 1, the customer finds the amount of maintenance unacceptable as there is too much disturbance to business operations. Scenario 2 shows how the capability for a certain level of availability has shifted from the maintenance service to the machine: here, the machine is re-designed to require less maintenance. The next step details on approaches to develop PSS designs based on the gaps identified in the last two steps.

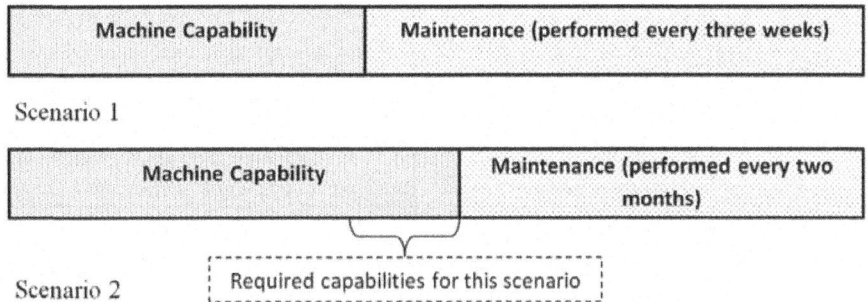

Figure 5: Scenarios 1 and 2: shifted availability division between product and service capability.

Step 4: identify new design required to satisfy needs at every level

The outcomes from steps 1 to 3: customer needs and capabilities and the properties of existing apposite products and services, will highlight the current gaps which could be filled by PSS designs. In particular, the capability map of the customer's processes will pinpoint the focus areas for development. Based on the conceptualization of PSS design through the network of seven parameters described in the 'What constitutes PSS design?' section, a new design should be generated for each gap by either introducing and/or modifying the current system in terms of activity, resource, competence and environmental influence. In this step, PSS designers have to identify possible designs for each gap identified. The designs should aim to eliminate or reduce the gap. For each gap, designs need to be explored at all of these levels. Design generation is a creative activity. To structure this process at the conceptual stage, we mapped resources for objects and competences for parameters and values. Table 2 illustrates sample designs generated for a gap in these elements. The subsequent sections illustrate steps in the framework to understand, evaluate, chose and represent the design.

Table 2: Illustration of usefulness of mapping elements in design generation for a gap

Gap	Mapping elements	Designs
Difficulty in cutting parts with varying complexity and different shapes	Objects (resources)	1. New system with inbuilt knowledge
		2. Modify feature extraction mechanism
	Activities	1. To develop a technique to group parts
		2. New technical assistance team
	Parameters and values (competences)	1. The customer should specify the cutting parameters and respective values
	Environment	1. Availability of using off-the-shelf standard parts

Step 5: identify conditions and consequences of each design

In this step, the conditions and consequences of each generated design from the last step need to be identified. This is an important step in understanding the ramifications of every generated design. The conditions should detail the circumstances which are required by the particular design. The consequences express the changes which occur in the system or which could affect another system if the particular design is executed. Since the overall system impacts upon both the conditions and consequences, these could be expressed in terms of environmental factors, product and service attributes, as well as economic and socio-cultural factors. At the conceptual stage, it can be difficult to identify the conditions and consequences. For this reason, the heuristic judgement of experts can play a vital role in predicting the conditions and consequences of each design; organizing a group exercise between experts would enrich the specification of conditions and consequences. From the conditions and consequences, the additional capabilities required for each design are derived in the next step.

Step 6: identify (additional) capabilities required to develop new designs

The identified designs that have been selected (given the understanding of the customer's capabilities) will help the provider to develop their own capabilities along with those of the supply network. The commonalities and differences between the stakeholders' capabilities need to be explicitly shared and understood by the stakeholders. The following processes are involved in explicating the capabilities required:

- For each design, a detailed list of activities to be carried out to implement the design should be documented.

- The evaluation of the required resources and their efficiency in each activity and which stakeholders are best placed to offer them.

- Identification of the parameters, which can be monitored and controlled by stakeholders, within each design considering past, present and future scenarios.

Shifts in the capabilities between the stakeholders which may then require additional resources should be carefully aligned and integrated. Any difficulties the stakeholders may have in meeting increased capability demands could be compensated for with variations in resources and time constraints. In the next step, the additional capabilities required are assigned between stakeholders.

Step 7: understand and specify the responsibilities of the stakeholders

Based on the required capabilities identified from the previous step and the preferences and views of stakeholders, responsibilities have to be assigned. Responsibilities could be taken at various levels: activity, object and parameter levels. These assignments form a core part of the PSS design process. This step develops various business models based on the responsibilities undertaken by the stakeholders. This alignment of responsibilities will precisely define network relationships. Various soft elements play vital roles in such relationship development such as trust, confidence, commitment and culture. The development of an open network would be valuable as the responsibility map should ideally be visible to all in the network, since, ultimately, all of the responsibilities are the concern of all in the network. In the next step, different PSS designs and responsibilities of the stakeholders across identified gaps are merged to meet customer needs.

Step 8: group different combinations of designs to satisfy customer needs

While analyzing product life cycle activities along with the customer business process, it is important to improve PSS at an overall system level and avoid sub-optimizing any individual activity. To avoid sub-optimizing any single activity and to satisfy the identified customer needs, different combinations of designs addressing different activities could be grouped based on the responsibilities assigned. Although synthesizing different designs could produce a variety of possible designs, this could expand exponentially and, therefore, could be difficult to manage. To systematically explore this process, Taguchi's factorial method [22] could be used to synthesize the different designs generated. Taguchi's method intends to ensure sufficient performance at the design stage of products and services. It is a systematic procedure in which all controllable factors (except one) are held constant as a variable factor is altered discretely. The controllable factor is considered to influence the simulation response, and its level can be controlled by designers. The control parameters identified in each design in step 6 could be used to vary the values to generate multiple options within each design. This synthesizing process also helps in understanding the sensitivity of important variables in the performance of the proposed PSS designs. At the conceptual stage, the detailed application by the development of Taguchi's orthogonal matrices may not be required. Each synthesized design should be evaluated in detail in the next step.

Step 9: evaluation of the cost and functional performance of the (grouped) designs

Evaluation should be part of every step in the proposed framework. To emphasize this evaluation process, it is dealt with separately in the framework. The evaluation should focus on three primary dimensions: economic, social and environmental [20]. From a business perspective, the major evaluation criteria will be profit, customer satisfaction, quality of products and services, value-in-use and risk reduction [23]. Both the tangible and intangible merits and demerits should be evaluated. To evaluate and choose the generated designs for an intended period, the sequence of activities involved in the grouped design has to be specified. By varying the conditions and consequences of designs, the responsibility assignments, and the combinations of designs for each gap, a variety of designs could be evaluated. To understand variable changes at important distinct points for an intended period, discrete event simulation is recommended at this stage to identify the performance of each design. In the final step, the evaluated and chosen design should be represented clearly to ensure a common interpretation across stakeholders.

Step 10: representation of the chosen design

The final step is to represent the chosen design in a format which could be easily generated and commonly interpreted by the stakeholders involved. The primary motivation in PSS modelling is to co-produce conceptual models that can be systematically shared by stakeholders. By analyzing various elements discussed from the above steps, a common representation map through extended IDEF0 modelling has been developed (Figure 2). The representation is based on the mapping of each activity in PSS design into inputs, outcomes, resources, competences, responsibilities, environmental variables and customer needs. The sequence of activities is linked as each activity is based on the satisfaction of preceding activity's outcomes. This representation emphasizes important PSS parameters and the interactions amongst them. This representation helps to highlight the current and modified system states, and assesses the performance of each activity involved in the grouped design. This representation appears to be simple, flexible and easy to maintain. Advantages of this framework and representation are illustrated using an industrial case study in the following section.

FRAMEWORK CORROBORATION WITH A LASER SYSTEM CASE STUDY AND IMPLEMENTED IN ISCL

To corroborate the proposed framework, a step-by-step application of the framework to laser systems for cutting operations which are used in manufacturing is used. A number of interviews were conducted with laser job shop owners (users of laser cutters), sales managers of manufacturers of laser cutters and also with experienced academics within this field to determine the current level of servitization within laser job shops. Table 3 provides details about the approaches used to build this laser system case study and for its evaluation. A step-by-step illustration to implement a higher level PSS is offered.

Table 3: Details about the approaches used to collect necessary data

Purpose	Roles and re-sponsibilities	Number of inter-views	Approaches	Duration
To build a laser system case study to apply the developed framework	Laser job shop owner (laser system user)	3	Semi-structured interviews - questions covered broad concerns of job shops: business needs, laser system specification, available resources, types of customers, applications, usage, current business solutions, types of services, laser system life cycle and laser cutting process parameters	Approximately 1 h each
	Laser system manufacturer	2	Semi-structured interviews - questions covered the broad topics of company background, types of laser system manufactured, types of customers, types of services, types of business solutions, supplier network and costing process	Approximately 1 h each
Evaluation of the developed laser case study along with the proposed framework	Academic researchers on laser system and processes	2	Discussion and feedback received through a Power-Point presentation	Approximately 1 h each
Evaluation of the framework with industrial practitioners	Product development team	3	Discussion and feedback received through a Power-Point presentation and also filled an assessment sheet to rate the proposed framework in a 5-point scale for potential usefulness, completeness, usability and clarity	Approximately 30 min to 1 h
	Business programme operation team	1		
	Technology development team	1		

To assist PSS designers computationally, the framework has been illustrated through ISCL [4]. ISCL has been chosen over other software proposed in the literature for the following reasons:

- ISCL is well aligned with our framework especially with regard to the elements used: entities in a service environment, attributes, specifications and activities changing attribute values and realizing specifications.

- ISCL supports life cycle simulation through a quantitative and probabilistic approach, which is important for assessing PSS designs. Life cycle simulation evaluates product life cycles from an integrated view of economic profitability and environmental awareness and optimizes the life cycles [6]. In ISCL, life cycle simulation is implemented through a discrete event simulation technique applied to life cycle design, such as the selection of end-of-life options (e.g. reuse, recycle and remanufacturing), the design of product modularity considering the options, and the timing and contents of service during the contract period.

- Standard process modelling and simulation approaches are not encouraged in this work because the PSS domain requires a specialized software environment defining its own terminologies and incorporating specific methods to support development. It helps to develop and integrate PSS research knowledge generation and understanding into a specific platform for wider uses and support tools evaluation.

- Although there is significant scope for improvements, from the authors' opinion, ISCL is a mature, reliable and accessible PSS software which well integrates modelling and simulation modules.

- Finally, the software is currently in the public domain and is actively supported by the developers, which is an incentive for industries to use it in practice.

Before describing the case study, the PSS modelling method employed in ISCL is briefly explained. A PSS-based business model in ISCL consists of activities, scenes and entities which are described along with their attributes (what the entity owns) and specifications (what the entity aims are) (Figure 6). Scenes represent partial states of the service environment as defined by the attribute values of entities. Activities treat scenes as the execution condition and change the value of attributes and realize specifications. As shown, these elements have several ports that are connected with lines. These lines have 11 different relations between elements instantiated during life cycle simulation. Figure 7 illustrates these links, and Table 4 shows the meanings (for details of the modelling method and grammar of the simulation codes, please refer to [24] and the user manual on the supporting website [25], respectively).

Designers can create these elements on the canvas of ISCL and also move, inspect and delete these elements on the canvas.

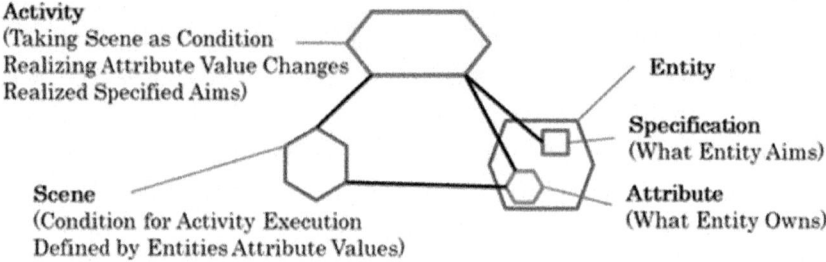

Figure 6: Elements in a business model on ISCL.

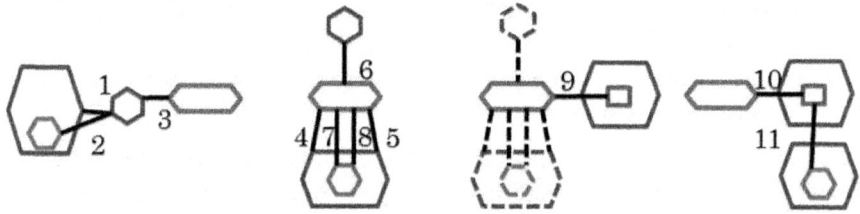

Figure 7: Relations in a business model on ISCL (please read Table 4 for link description).

Table 4: Definition of relations in a business model on ISCL in relation to Figure 7

Figure 7 link number	Element 1	Element 2	Meaning
1	Scene	Entity	Scene includes all instances of Entity
2	Scene	Attribute	Entity instances included in Scene are specified by the value of Attribute
3	Activity	Scene	Scene is regarded as the execution condition of Activity
4	Activity	Entity	Activity creates instances of Entity
5	Activity	Entity	Activity deletes instances of Entity
6	Activity	Scene	Activity refers to (calls) an instance of Entity specified by Scene
7	Activity	Attribute	Activity gets the value of Attribute
8	Activity	Attribute	Activity assigns the value of Attribute
9	Activity	Specification	Specification is realized as a result of execution of Activity
10	Activity	Specification	Specification is evaluated during the execution of Activity
11	Specification	Attribute	The value of Specification is related with Attribute

Step 1: understanding customer needs

The laser systems under consideration in this case study are mature products as are the laser processes which are structured and mostly inbuilt to the system. The customers are laser job shop owners who procure laser systems from the original equipment manufacturer and supply semi-finished goods to the end product manufacturer. Figure 8 illustrates stakeholders' map of the laser system case study. As the laser job shops have many years of experience in this field, they are able to precisely specify their requirements of laser-cutting systems. The transaction type between the laser system manufacturer and the laser job shop customer is business-to-business. The semi-structured interviews with the laser job shops revealed the importance of the need behind the need, experiences, required state change and business ambitions of the laser job shop. We developed a PSS customer needs specifications of the laser system by using overall equipment effectiveness which is a multiplication of availability, performance and quality. The required values are mapped to be:

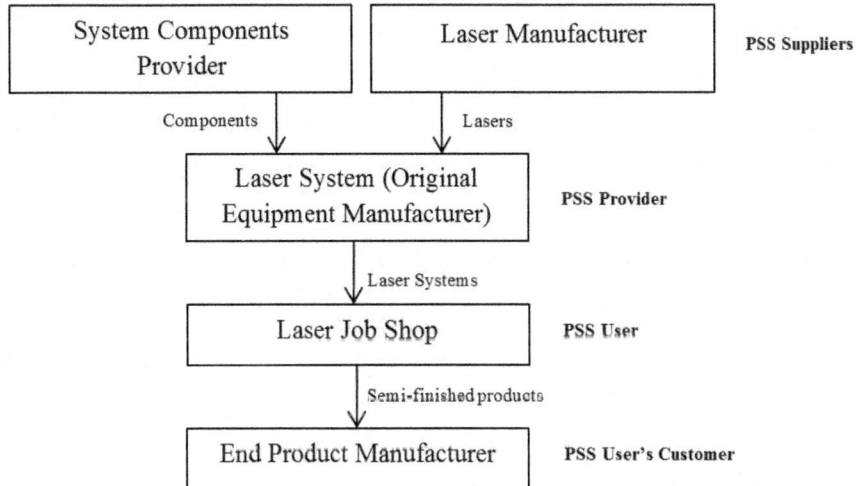

Figure 8: Stakeholders' map of the laser system case study.

- Overall equipment effectiveness ($A \times P \times Q$), 50% to 60%
- Availability (Mean time between failure (MTBF) / (MTBF + Mean time to repair (MTTR)), 85% to 95%
- Performance (Working speed / Designed speed), 55% to 65%
- Quality (Good units within tolerance / Total units produced), 95% to 99%
- Laser system usage period, 5 years

At this point, ISCL supports the designer to define the specifications of the users of laser systems as well as the attributes of the laser systems, which are identified at this step. Figure 9 details the specifications and the attributes of the laser systems with examples. In Figure 9a, three entities: 'Manufacturer', 'LaserSystem' and 'User', are shown. The model includes an activity 'Use' to deliver 'Function' as a specification targeted by User. LaserSystem already includes the overall equipment effectiveness specification and the relevant attributes such as *MTBF* and *MTTR*. The value of these attributes can be individually calculated with respect to each instance of LaserSystem during the life cycle simulation. Figure 9b shows the objectives of the model, which are statistical values obtained as a result of life cycle simulation. For instance, 'AvrOEE' is the average overall equipment effectiveness of all laser systems in the market with respect to simulation time, which is defined with the window in Figure 9c. At this moment, the dynamic behaviour of laser systems such as physical deterioration has not been defined yet. In the next step, to understand whether required needs levels are achievable through existing customer's capabilities, they are noted.

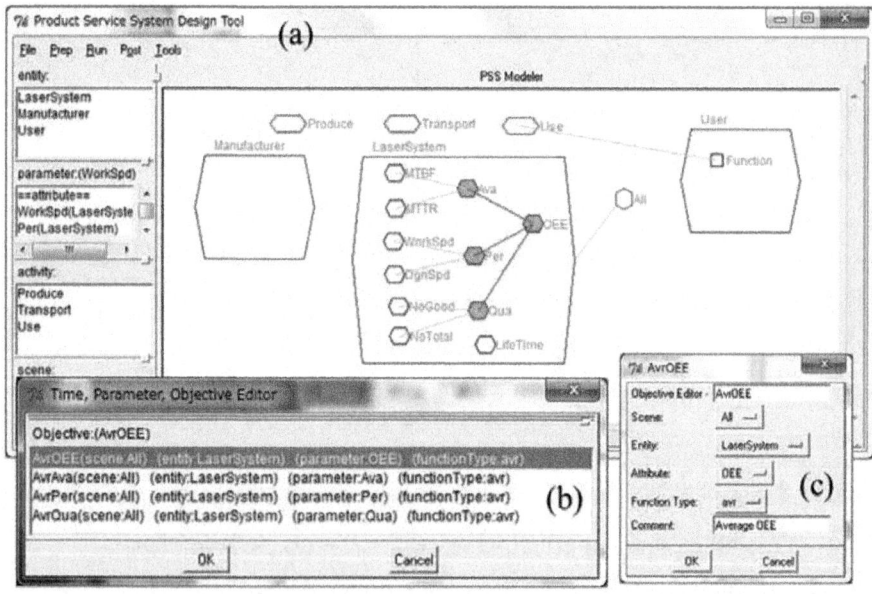

Figure 9: Modelled attributes of laser systems and their objectives in ISCL. (a) Entities of the model. (b) Objectives of the model. (c)Average overall equipment effectiveness.

Step 2: identification of the existing capabilities and resources of the laser job shop

To understand the laser job shop's capabilities, currently performed activities are mapped with resources, outcomes and the reasons for deficiency. The important tasks which are mapped include machine calibration, CAD file preparation, identification of process parameters, material preparation, work piece (un)loading, work piece alignment, machining operation, daily and planned maintenance, repair work, and material and gas procurement. Some of the activities to map the laser job shop's capabilities are detailed in Table 5. The complete list of activities and the respective status of each helps us to understand the capability gaps of laser job shops. Along with these gaps, current products and services are assessed in the next step.

Table 5: Mapping existing capabilities of laser job shops

Activities	Resources	Outcomes	Reasons
Machine calibration	Limited skills. One employee	Trial and error	Not having enough understanding of the machine. The system is partially protected by the manufacturer
Work piece loading/unloading	Manual process. Two unskilled employees	Time-consuming	Automated machine unavailable

Step 3: identify current status of laser systems and services

A laser system is an assemblage of a laser generator unit, beam delivery system, beam manipulation system, motion system, process monitoring system and a control system (Figure 10). The key performance indicators (KPIs) for these systems are failure rate, repair time, degradation rate, redundancy and reliability. Mapping these KPIs to the sub-systems shows that beam delivery and beam manipulation systems have to be improved. Identifying root causes through discussions with the experts revealed that mirror misalignment, laser instability, variation within suppliers, operator's error and a mismatch in cooling needs could be possible problems to be addressed. Mapping the KPIs of services (frequency, number of technicians available, time consumed, spare parts and tools availability, and location) with provided services (training, planned maintenance, technical assistance and repair activity) revealed several potential improvement areas like the operator's knowledge of the machine, complexity regarding the different shapes to be machined, constraints in space requirements, the probability of making mistakes being high and escalated expense of module replacement. In this study, there was restriction to collect required industrial data. Therefore, the current scenario was simulated through data collected from the interviews. If data could have been available, ISCL supports to import these data through .CSV format. From the interviews, it was revealed that the overall equipment effectiveness at the required level is not maintained and, in particular, that performance should be improved.

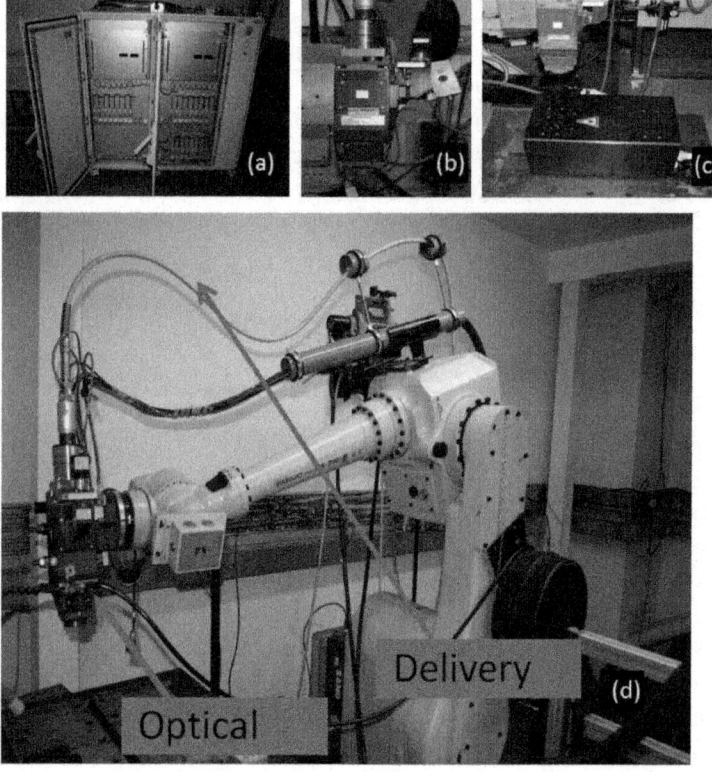

Figure 10: Laser system. (a) Laser unit (IPG YLR-8000). **(b)** Optical head. **(c)** Monitoring system. **(d)** Robot motion system.

At this point, ISCL supports the designer in adding other necessary elements to describe the current design. The refined model (Figure 11) includes new activities, entities and attributes so that the model can represent the current design. For instance, 'Fail(ure)Rate', 'Deg(radation)Rate' and 'Rep(air)Rate' are added as attributes of LaserSystem, which influence the degradation in terms of the availability, performance and quality. Furthermore, 'Engineer' and 'Operator' are treated as the entities of the current design. Their 'Skill' also influences the degradation and its recovery through activities 'Repair' and 'Maintenance'. By supplying codes specific to each activity, the model becomes the input of life cycle simulation. For instance, Figure 12a shows the codes specific to an activity Repair. These codes are used to automatically create links between the model elements as shown in Figure 12b, which is helpful to the designer in debugging these codes. An explanation of coding as the input to life cycle simulation is beyond the scope of this paper. Figure 13 shows the simulation result of the current design, the average overall equipment effectiveness with respect to simulation time. At this stage, the simulation results are partly based on fictive parameter values, which should be specified in the design process. The gaps identified in these first three steps facilitate generation of PSS designs in steps as described in the following sections.

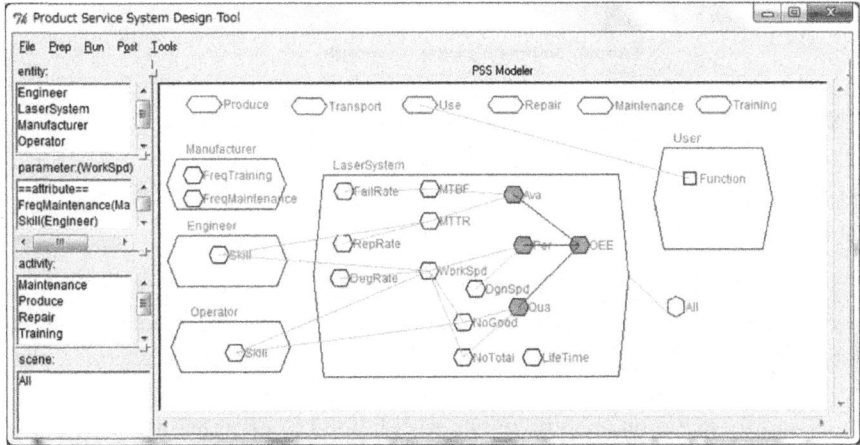

Figure 11: Modelling the current design in ISCL.

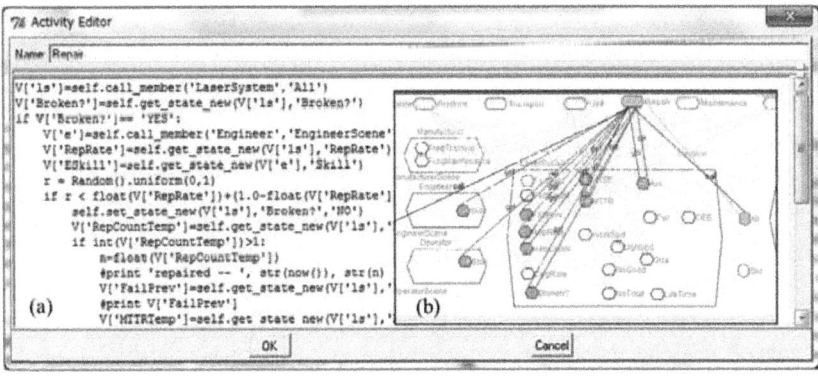

Figure 12: The detail of an activity repair in the current design. (a) Codes specific to an activity Repair. (b) Links between the model elements.

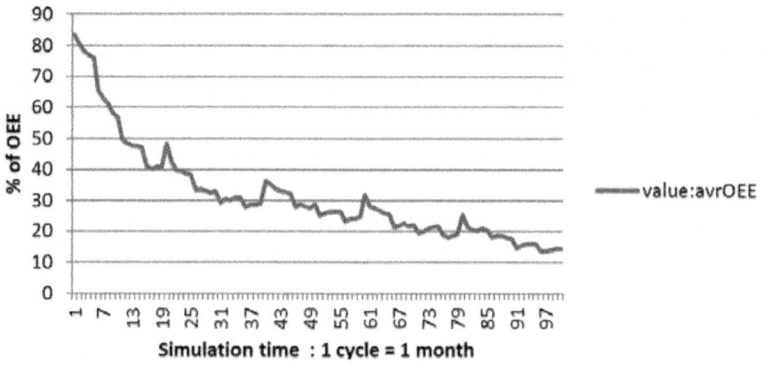

Figure 13: The simulation results of the current situation.

Step 4: identification of new designs required to satisfy the needs at every level

From steps 2 to 3, 10 activities have been identified for improvement. Within the 10 activities, 18 problems are observed. Through group brainstorming with the researchers, 54 designs have been generated. From the perspectives of adding and modifying activities, enriching the laser systems through support systems and focusing on specific parameters helped to generate many designs.

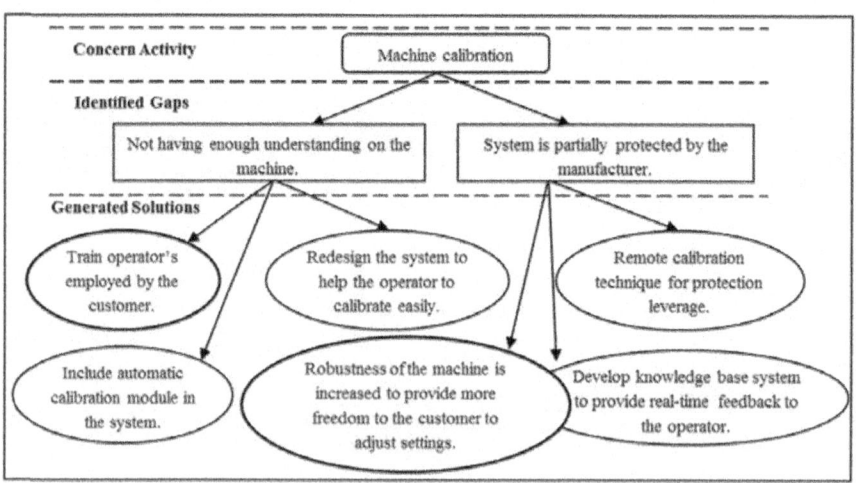

Figure 14: Representation of generated designs for two problems in the concern activity.

These designs have been checked and rated by the researchers in preference with the feasibility scope. The highlighted circles in Figure 14 are some of the preferred designs. In order to add these designs in ISCL, corresponding model elements should be added to the model to represent the current design. Alternatively, corresponding model elements can be separately modelled and stored in the knowledge base of ISCL. The stored model elements are instantiated when necessary. Furthermore, the formalization of this step can be useful for ISCL to automatically synthesize new designs using the knowledge base. The subsequent steps help to develop and assess these designs.

Step 5: identify conditions and consequences of each design

Steps 5 and 6 are jointly discussed in the next section due to greater continuity between them.

Step 6: identify (additional) capabilities required to develop new designs

Even though preferred designs are highlighted in the previous step, the potential of every design is identified by specifying conditions and consequences. The conditions are mainly specified through the frequency and complexity of the activities, laser system usage period and patterns, required skills, parameters values, man-days required and the condition of the laser system. The consequences of each of the proposed designs are expressed in terms of the impact on availability, performance and quality to set the overall equipment effectiveness and incurred costs. The generated conditions and consequences have been evaluated by the heuristic judgements of two experts. From the conditions and consequences, the capabilities required by each design are derived. Table 6 illustrates the conditions, consequences, capabilities, resources and control variables derived for new designs which have been generated. Conditions and consequences are inherent processes in ISCL. Support should be developed to highlight the new capabilities achieved, and the additional resources should be incorporated in the PSS laser system modelling. Developing new integrated business models between the laser job shop, the manufacturer and suppliers through the alignment of capabilities for each design will be the next step.

Table 6: Outcomes from steps 5 and 6 for the derived new design

Solutions	Conditions	Consequences at each time	Capability shift	Resources	Control variables
Technical assistance to set efficient process parameters	Laser parts (new shape and size parts)	Man-days, +0.2 (24-h day); performance, +0.5%; quality, +1%; set (overall equipment effectiveness (laser system)); total cost, +£50	Efficient process parameters irrespective of varying shapes and sizes	To employ an additional three technical employees to support process query	Frequency of new shapes and sizes, technical employees, number of queries

Step 7: understand and specify responsibilities of the stakeholders

Responsibilities have to be assigned precisely between the laser job shop, the manufacturer and suppliers (based on the identified required resources) for the capability shift to occur to satisfy the proposed design and eventually to satisfy the required level of overall equipment effectiveness. These assignments are precisely defined either at the level of activities, objects or parameters. Based on inputs of preferences and views from the laser job shops and manufacturers, various possibilities of responsibility assignment are generated. The outcome of such a PSS design process has stark impact upon business model generation. Table 7 illustrates responsibilities aligned within a new design generated. Using ISCL, such responsibilities are defined by selecting appropriate entities (such as Manufacturer) as the service provider of specific activities. This assignment results in modification of financial flow. In some cases, the expected service may not be realized in life cycle simulation, because some entities may not satisfy the conditions to provide the expected service (such as the level of Skill of Engineer). Synthesizing designs generated for 18 problems through responsibility alignment is the next step.

Table 7: Responsibility assignment at various levels for new designs

Responsibilities	Laser job shop	Manufacturer	Supplier
Activities	Identifying machined part variety	Measurement of operator skills	To develop learning content
Objects (resources)	Experienced operators	Web-based support system	Control system for error identification
Parameters (competences)	Skills of operators	Frequency of training	Queries redirection

Step 8: group different combinations of designs to satisfy customer needs

To satisfy the required overall equipment effectiveness levels throughout the usage period, a wide variety of designs are generated by synthesizing the designs identified for each gap. The synthesis process is systematically carried out through Taguchi's factorial method. The control variables identified in each design in step 6 are used to vary values to generate multiple options within each design. Table 8 details the combinations possible across four activities. In Table 8, the four columns explain the chosen four activities (in which gaps exist in the current system) to be addressed in PSS designs. The six cells in each column explain two solutions with three improvement levels each (based on Taguchi's factorial method) to fill the gap in each activity.

Table 8: Combination of designs possible for four activities of focus and variable change in brackets

	Chosen four activities in which gaps exist in the current system			
	Machine calibration	Process path/ parameters	Daily maintenance	Repair work
Two solutions with three improvement levels each (control variable in bracket)	S1-CalibrationTraining-1 *(operator's skill level)*	S3-GroupTech-1 *(system's reliability)*	S5-OperTraining-1 *(operator's skill level)*	S7-DiagnosSys-1 *(system's reliability)*
	S1-CalibrationTraining-2	S3-GroupTech-2	S5-OperTraining-2	S7-DiagnosSys-2
	S1-CalibrationTraining-3	S3-GroupTech-3	S5-OperTraining-3	S7-DiagnosSys-3
	S2-Calibrateassist-1 *(service level)*	S4-TechAsst-1 *(service level)*	S6-SysRedesign-1 *(system's reliability)*	S8-MTTRincres-1 *(time to repair)*
	S2-Calibrateassist-2	S4-TechAsst-2	S6-SysRedesign-2	S8-MTTRincres-2
	S2-Calibrateassist-3	S4-TechAsst-3	S6-SysRedesign-3	S8-MTTRincres-3

Similar to step 4, the automatic synthesis of the new design by combining model elements stored in the knowledge base including these activities can be implemented in ISCL. Currently, ISCL supports the automatic synthesis process, when the conditions and consequences of added activities are defined and these activities solely influence the state of a single entity with a hierarchical structure. For instance, [4] shows a procedure implemented in ISCL to generate possible functional upgrading services combined with repair services for medical equipment systems. The next step evaluates combinations to understand the satisfaction of required overall equipment effectiveness levels.

Step 9: evaluate cost and functional performance of the (grouped) designs

The primary questions to be answered in the evaluation process for the laser system case study are as follows:

- Which designs and combination of designs satisfy the required level of overall equipment effectiveness for a specific period?
- How much does each successful implemented design cost for the specific period?
- How should cost be shared between the laser job shop, manufacturer and suppliers for each design based on responsibility assignment?

Answering these questions provides predictable costs, cost transparency and maximal security that are important factors to be considered in laser business model selection. ISCL provides an exceptional environment to carry out life cycle simulation. Discrete event simulation is carried out by sequencing activities involved in the grouped designs. By varying possible conditions and consequences of designs, responsibility assignments and combination of designs for each gap, a variety of designs are evaluated. Figure 15 points out the increase in overall equipment effectiveness levels for each design identified to improve machine calibration activity. All of the identified 54 designs are evaluated, and the best designs for each gap are synthesized to achieve the aforementioned overall equipment effectiveness intervals. The final synthesized chosen design should prove to be an improvement of operator skills, module repair, as well as daily, preventative and repair maintenance by advanced and frequent training, an efficient diagnostic and prognostic system and enhanced support systems. Figure 16shows that the overall equipment effectiveness depreciation of this combined design would satisfy the required value throughout 5 years. The cost sharing between the laser job shop, manufacturer and suppliers is shown in Figure 17. The representation of this chosen design is detailed in the last step.

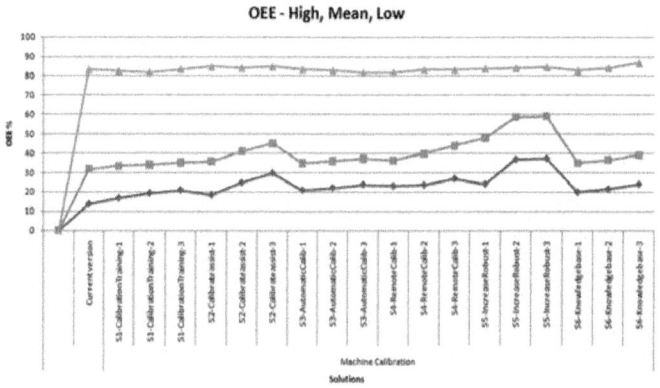

Figure 15: Overall equipment effectiveness increases due to the implementation of the new design (for three improvement levels each).

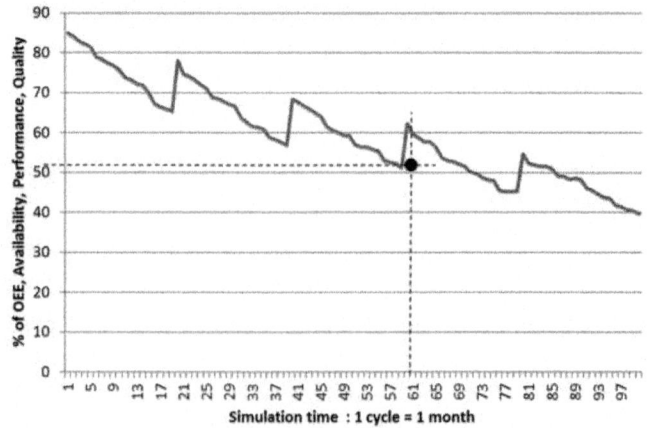

Figure 16: Overall equipment effectiveness depreciation for the proposed finalized design. The black dot represents the required level of overall equipment effectiveness to be maintained.

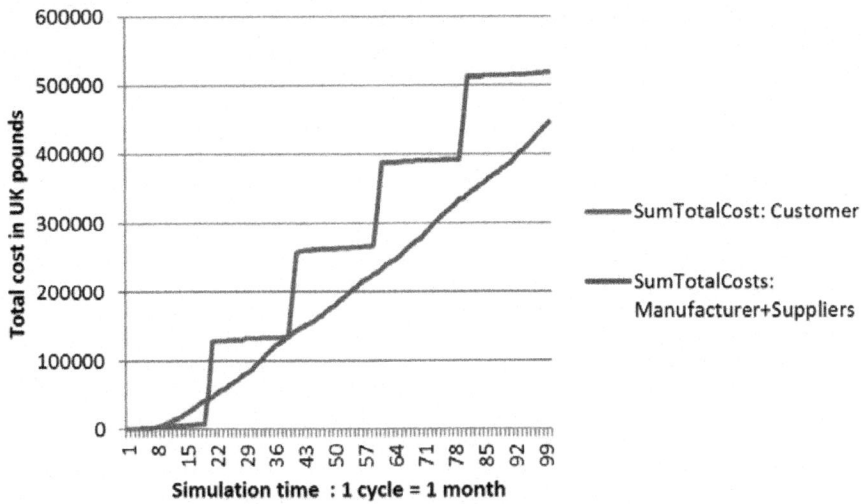

Figure 17: Cost sharing between the laser job shop, manufacturer and suppliers.

Step 10: representation of chosen design

The chosen design is represented by an extended IDEF0 format for common interpretation amongst the stakeholders involved (Figure 18). It highlights the activities of repair, maintenance, replacement module, daily maintenance, operator training and preventive maintenance along with customer needs, environmental variables, responsibilities, resources, the current and reached states, outcomes of a particular activity and interactions amongst them. The modifications to the current system are highlighted through dotted boxes (reduced daily maintenance) and colour changes. It represents alternatives between activities (repair maintenance and module replacement) and achievable outcomes. A module should be developed in ISCL to implement the extended IDEF0 representation. Also, this representation helps to frame the final contract using terms and conditions that are relevant to all of the stakeholders involved. The academic and industrial views of these steps and the results produced are detailed in the next step.

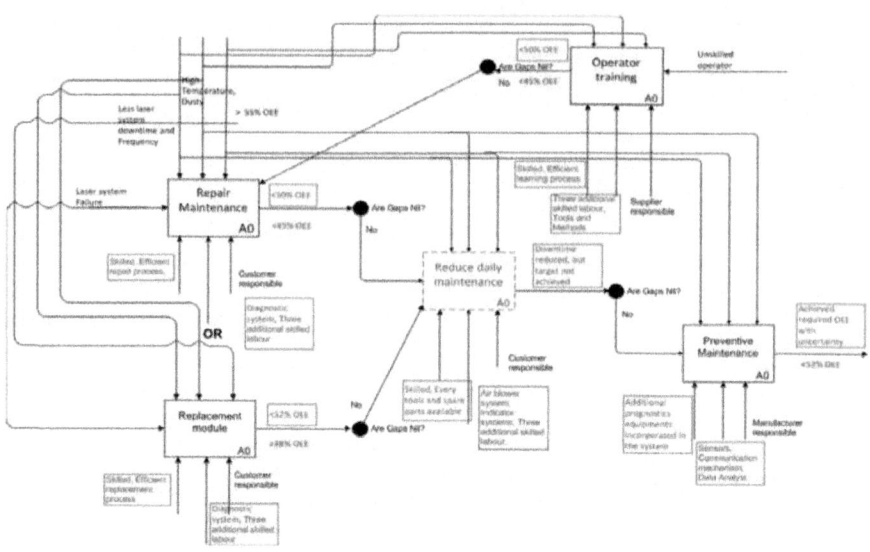

Figure 18: Extended IDEF0 model of the synthesized chosen PSS design.

ACADEMIC AND INDUSTRIAL VIEWS

To understand merits and limitations of the proposed framework, it has been corroborated by a step-by-step application with two laser system experienced academics and five industrial experts who are knowledgeable in PSS. The overall feedback is encouraging and potential applications of this framework have been stressed; notable benefits are that the structure and flow of the framework are clear and robust, the framework helps to develop PSS design from a holistic systems approach which facilitates a change in the designer's mindset from product-centric to systems-centric, the framework could be used as a general problem solving approach which incorporates the co-production process between relevant stakeholders, and the IDEF0 representation is perceived to be clear and informative. Potential usefulness, completeness, usability and clarity are highly graded by the experts (Figure 19). The limitations that have been expressed are that the level of trust and transparency required for this framework is absent in most industrial sectors, there exists a need to measure intangible benefits and to perform more quantitative data analysis to assess gaps, and the assessment of the skill set of industries and the development of tutorials to train employees to apply this framework along with obtaining the necessary information required for each step could all be problematic.

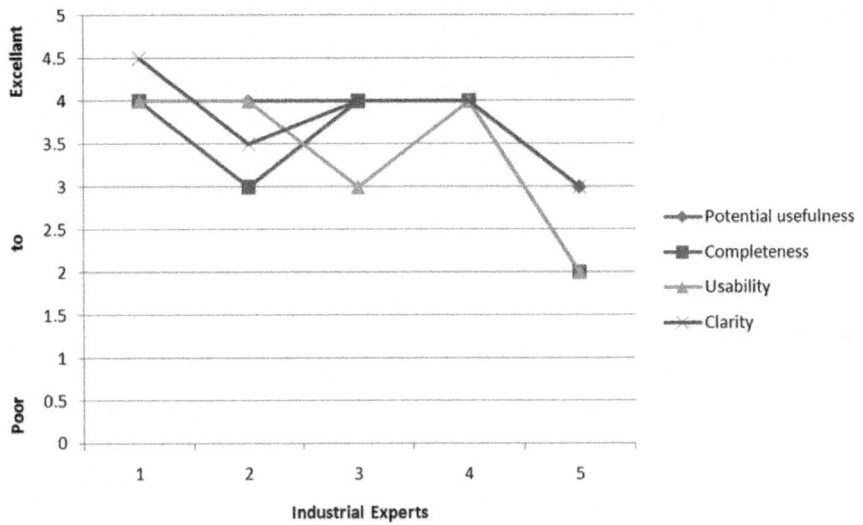

Figure 19: Five industrial experts' ratings for the proposed framework.

All of the discussed methodologies have novel steps in designing PSS. However, compared to other methodologies, the merits of the proposed framework are that its starting point is an existing system and that the customer needs are elicited as deficiencies within that system, it develops PSS designs which are aligned to the stakeholders' capabilities, and it synthesizes solutions to produce optimized designs which can encompass the whole of the customer business process. We believe that the proposed framework could be further developed by applying it to industrial case studies to develop functional and sustained PSS designs. As most of the proposed methodologies in the literature have not been formally applied to industrial case studies and demonstrated, a comparison using performance metrics is not possible. Further work in validation would involve a methodological effectiveness comparison to measure performance metrics. A list of performance metrics that could be potentially applied in this comparative exercise are tabulated in Table 9.

Table 9: PSS design methodology performance metrics parameters

	Performance metrics parameters
Methodological applicability	Potential usefulness
	Completeness
	Usability
	Clarity
	Systematic approach
	Reliability
	Effectiveness of feedback loops
	Change in designer's perspective from product-centric to PSS-centric
	Avoidance of re-work
	Time consumed in designing and delivering PSS
Methodological outcomes	PSS design functional performance for a specified duration
	Customer satisfaction (requirements and solutions)
	Enhanced value-in-use
	Agility and responsiveness of designed PSS
	Reliability of PSS
	PSS revenue and profit generation
	Whole life cycle PSS design
	Optimization of costs and resources
	Effectiveness of integrated system and co-production
	Common interpretation of PSS
	Innovative and usable PSS designs
	Minimum number of loopholes in PSS
	Robust stakeholders network and responsibility adherence
	Sustainable PSS solutions by stakeholders' capabilities
	Quality of PSS
	Prolonged relationship with customers
	Product/service improvements
	Understanding and control of the existing and developed system
	Validity of simulation results in real time
	Diffusion of product and service integration

DISCUSSION AND CONCLUSIONS

Based on identified gaps from the literature, a capability-based PSS design framework is proposed. This framework highlights the important features required in designing PSS to help meet customer goals. A capability-based approach generates a wide variety of PSS designs which are intended to produce sustained functional behaviour in the proposed system to achieve its purpose continuously and create an innovative value addition for the customer.

It encourages a broader customer business perspective in designing PSS as compared to the traditional approach where the functional availability of products is considered given their deterioration during the use stage. Details of a capability map of the customer's processes help to pinpoint the focus areas for PSS development. It stresses the capability shift and mix between products and services and also between stakeholders in order to develop feasible and enduring PSS designs.

The framework emphasizes the importance of taking into account the preferences and views of the stakeholders and the apposite alignment of responsibilities. Organized feedback loops between the various steps aid PSS design modelling and performance evaluation. The framework facilitates the exploration of a wide variety of designs by synthesizing designs that have been identified for various gaps. The generation and evaluation of designs are well structured in the framework. The results of the simulation using ISCL provide good insights into PSS designs. We believe that this framework would facilitate and structure the interactions between the customer, manufacturer and supplier. It also helps in understanding the capabilities of the stakeholders and aids an understanding of the value of PSS designs as appreciated by the customer. The representation of PSS designs through extended IDEF0 was found to be useful in providing a common understanding among the stakeholders.

The application of this framework to a real industrial laser system case study helps to demonstrate the benefits of the framework. This demonstration also helps refine the framework based on the few observed limitations such as the inputs collected and the features in the modelling technique. The demonstration of the framework along with ISCL highlights the important features of ISCL and the modules to be developed to enhance design support. In particular, support for design generation through standardized PSS ontology and the synthesis of different designs generated through Taguchi's factorial method would prove to be especially helpful for PSS designers. Additionally, the implementation of this framework using a step-by-step approach in software could greatly facilitate the design process by decreasing information load on PSS designers. An effective mechanism is required to support PSS designers to specify the conditions and consequences of each development design which are vital inputs for PSS simulation. Since the underpinning expectation that PSS will have a lower environmental impact is yet to be proven, the impact of PSS designs on environmental benefits should be studied separately. There is ongoing work in applying this framework to various case studies which involve other case companies who are in the process of refining the development of their PSS designs.

Declarations

ACKNOWLEDGEMENTS

This research was funded by the Cranfield Innovative Manufacturing Research Centre (EPSRC Grant EP/E001874/1) as part of the research project 'Capability based conceptual PSS design'.

Authors' contributions

GVAV carried out understanding industrial challenges, framework development, case study development, framework evaluation and manuscript preparation. HK participated in the framework development, Service CAD integration with the framework and manuscript preparation. RH participated in understanding industrial challenges and assisted in the framework development, case study development and framework evaluation. RR was the principal investigator of this project and assisted in all the phases of the work especially on the framework development and framework evaluation. TT supported in Service CAD integration with the framework. SE and AT participated in the framework development and framework evaluation. SW supported in the laser case study development and framework evaluation. All authors read and approved the final manuscript.

REFERENCES

1.   Neely A: Exploring the financial consequences of the servitization of manufacturing. *Operation Management Research* 2008, 1:103–118. 10.1007/s12063-009-0015-5

2.   Annamalai Vasantha GV, Roy R, Lelah A, Brissaud D: A review of product–service systems design methodologies. *J. Eng. Des.*2012,23(9):635–659. 10.1080/09544828.2011.639712

3.   Mont OK, Plepys A: *Customer satisfaction: review of literature and application to the product-service systems. Final report to the Society for Non-Traditional Technology, Japan.* International Institute for Industrial Environmental Economics, Lund University, Lund; 2003.

4.   Komoto H, Tomiyama T: Integration of a service CAD and a life cycle simulator. *CIRP Ann. Manuf. Technol.* 2008, 57: 9–12. 10.1016/j. cirp.2008.03.001

5.   Komoto H, Tomiyama T: Design of competitive maintenance service for durable and capital goods using life cycle simulation. *Int. J. Automot. Techn.* 2009,3(1):63–70.

6.   Umeda Y, Nonomura A, Tomiyama T: Study on life-cycle design for the post mass production paradigm. *Artificial Intelligence for Engineering*

Design, Analysis and Manufacturing 2000, 14: 149–61. 10.1017/S0890060400142040

7. Sakao T, Shimomura Y: Service engineering: a novel engineering discipline for producers to increase value combining service and product. *J. Clean. Prod.* 2007,15(6):590–604. 10.1016/j.jclepro.2006.05.015

8. Shimomura Y, Hara T, Arai T: A service evaluation method using mathematical methodologies. *CIRP Ann. Manuf. Technol.*2008,57(1):437–440. 10.1016/j.cirp.2008.03.012

9. Shimomura Y, Hara T, Arai T: A unified representation scheme for effective PSS development. *CIRP Ann. Manuf. Technol.*2009,58(1):379–382. 10.1016/j.cirp.2009.03.025

10. Maussang N, Zwolinski P, Brissaud D: Product–service system design methodology: from the PSS architecture design to the products specifications. *J. Eng. Des.* 2009,20(4):349–366. 10.1080/09544820903149313

11. Alonso-Rasgado T, Thompson G, Elfström B: The design of functional (total care) products. *J. Eng. Des.* 2004,15(6):515–540. 10.1080/09544820412331271176

12. Alonso-Rasgado T, Thompson G: A rapid design process for total care product creation. *J. Eng. Des.* 2006,17(6):509–531. 10.1080/09544820600750579

13. Zhao Y, Harrison A, Roy R, Mehnen J: Aircraft engine component deterioration and life cycle cost estimation. In *Globalized Solutions for Sustainability in Manufacturing: Proceedings of the 18th CIRP International Conference on Life Cycle Engineering*. Edited by: Hesselbach J, Herrmann C. Berlin: Springer; 2011:657–662.

14. Neaga EI, Henshaw M, Yue Y: The influence of the concept of capability-based management on the development of the systems engineering discipline. *Proceedings of the 7th Annual Conference on Systems Engineering Research (CSER 2009), Loughborough University, Loughborough, 20–23* 2009.

15. Matzen D: *A systematic approach to service oriented product development.* PhD dissertation: DTU Management Engineering, Technical University of Denmark; 2009.

16. Tan AR: *Service-oriented product development strategies. PhD thesis, Department of Management Engineering, Technical University of Denmark.* 2010.

17. Hussain R, Lockett H, Annamalai Vasantha GV: A framework to inform PSS conceptual design by using system-in-use data.*Computers*

in Industry, Special Issue: Product Service System Engineering: From Theory to Industrial Applications 2012, 63: 319–327.

18. Tucker A, Tischner U: *New Business for Old Europe - Product-Service Development*. Greanleaf, Sheffield: Competiveness and Sustainability; 2006.

19. NASA: *NASA Systems Engineering Handbook: NASA/SP-2007-6105 Rev1*. NASA, Washington, D.C; 2007.

20. Annamalai Vasantha GV, Hussain R, Cakkol M, Roy R, Evans S, Tiwari A: An ontology for product-service systems. In *Decision Engineering Report Series*. Edited by: Xu Y. Cranfield: Cranfield University; 2010.

21. Annamalai Vasantha GV, Roy R, Cakkol M: Problem definition in designing product service systems. *Proceedings of the 3rd CIRP Conference on IPS2, Braunschweig, 5–6 May 2011*

22. Roy R: *A Primer on the Taguchi Method. Society of Manufacturing Engineers, Dearborn*.

23. *Supply-Chain Council: Supply-Chain Operations Reference-Model: SCOR Version 7.0 Overview. Supply-Chain Council, Cypress*. 2005.

24. Komoto H, Tomiyama T: Systematic generation of PSS-concepts using a service CAD tool. In *Introduction of Product/Service-System Design*. Edited by: Sakao T, Lindahl M. Berlin: Springer; 2010:71–91.

25. *ISCL software*. 2005. (2011). Accessed 20 Aug 2013 http://staff.aist.go.jp/h.komoto/iscl_en.html

CITATION

CHAPTER 1

Singh, R. , Singh, R. and Khan, B. (2015) A Critical Review of Machine Loading Problem in Flexible Manufacturing System. World Journal of Engineering and Technology, 3, 271-290. doi: 10.4236/wjet.2015.34028.

CHAPTER 2

Xiaomeng Li, Xiaoqing Wu, Peng Shi and Zuo-Guang Ye, Lead-Free Piezoelectric Diaphragm Biosensors Based on Micro-Machining Technology and Chemical Solution Deposition, doi:10.3390/s16010069.

CHAPTER 3

Qiaokang Liang, Dan Zhang, Gianmarc Coppola, Jianxu Mao, Wei Sun, Yaonan Wang and Yunjian Ge, Design and Analysis of a Sensor System for Cutting Force Measurement in Machining Processes, doi:10.3390/s16010070.

CHAPTER 4

Rosario Domingo, Marta María Marín, Juan Claver and Roque Calvo, Selection of Cutting Inserts in Dry Machining for Reducing Energy Consumption and CO_2 Emissions, doi:10.3390/en81112362.

CHAPTER 5

Wan-Jui Shen, Ming-Hung Tsai and Jien-Wei Yeh, Machining Performance of Sputter-Deposited (Al0.34Cr0.22Nb0.11Si0.11Ti0.22)50N50 High-Entropy Nitride Coatings, doi:10.3390/coatings5030312.

CHAPTER 6

L. Rihakova and H. Chmelickova, "Laser Micromachining of Glass, Silicon, and Ceramics," Advances in Materials Science and Engineering, vol. 2015, Article ID 584952, 6 pages, 2015. doi:10.1155/2015/584952.

CHAPTER 7

Wei Feng, Bin Yao, BinQiang Chen, DongSheng Zhang, XiangLei Zhang, and ZhiHuang Shen, "Modeling and Simulation of Process-Machine Interaction in Grinding of Cemented Carbide Indexable Inserts," Shock and Vibration, vol. 2015, Article ID 508181, 8 pages, 2015. doi:10.1155/2015/508181.

CHAPTER 8

Mi Xiao, Long Wen, Xi Li, and Liang Gao, "Modeling of the Feed-Motor Transient Current in End Milling by Using Varying-Coefficient Model," Mathematical Problems in Engineering, vol. 2015, Article ID 103507, 9 pages, 2015. doi:10.1155/2015/103507.

CHAPTER 9

Agnieszka Dmowska, Bogdan Nowicki, and Anna Podolak-Lejtas, "Surface Layer Properties after Successive EDM or EDA and Then Superficial Roto-Peen Machining," Advances in Tribology, vol. 2012, Article ID 723919, 12 pages, 2012. doi:10.1155/2012/723919.

CHAPTER 10

Vivek Jain, Apurbba Kumar Sharma, and Pradeep Kumar, "Recent Developments and Research Issues in Microultrasonic Machining," ISRN Mechanical Engineering, vol. 2011, Article ID 413231, 15 pages, 2011. doi:10.5402/2011/413231.

CHAPTER 11

S. Belmokhtar, A.I. Bratcu and A. Dolgui (2006). Modular Machining Line Design and Reconfiguration: Some Optimization Methods, Manufacturing the Future, Vedran Kordic, Aleksandar Lazinica and Munir Merdan (Ed.), ISBN: 3-86611-198-3, InTech, DOI: 10.5772/5047.

CHAPTER 12

Vasantha et al.: A manufacturing framework for capability-based product-service systems design. Journal of Remanufacturing 2013 3:8. doi:10.1186/2210-4690-3-8

INDEX